DESCARTES' DAUGHTERS:
THINKING-MACHINES AND THE EMERGENCE
OF POSTHUMAN COMPLEXITY

By

Dawn E. Lausa
B.A. Marietta College, 1984
M.A. Syracuse University, 1993

DISSERTATION

Submitted in partial fulfillment of the requirements for the
Degree of Doctor of Philosophy in English
in the Graduate School of Syracuse University

February 2009

Approved_____
Professor Gregg Lambert

Date_____

Descartes' Daughters:
Thinking-Machines and the Emergence of Posthuman Complexity

ABSTRACT

The very nature of the machine seems to signify determinacy, to preclude it from calling upon the kind of indeterminate creativity that characterizes human thought and reason. Indeed, at least since Descartes, the machine and the human have each been conceived precisely in opposition to one another, and in this sense, each is strictly inconceivable without the other. But, already within Descartes, the concept of contingency, expressed as speech, will, reason, and soul, thus folded into its own material actualizations, became the *function* of contingency. And this function, once sundered from substance, could be infinitely iterated both within and beyond the human. Thus, each iteration of the thinking-machine redistributes the component concepts of the human and the machine across the boundary of their respective identities, while transforming and destabilizing our concepts of both.

Chapter One begins by examining Descartes' concept of the human, and the significance of its identification in opposition to the machine, arguing that the effect of Descartes' multiple iterations of these oppositional concepts is to render the relationships among their components contingent rather than essential, and to transform the surrounding discourse of substance ontology to one of becoming function. Chapter Two examines the calculating engines of Charles Babbage as material iterations of the thinking-machine, and the discursive formations surrounding them, focusing upon an emerging functional image of thought. Chapter Three considers the ontological uncertainties that the thinking-machine provokes as a source of anxiety surrounding Maelzel's automaton chess-player, as evidenced in numerous texts, including Edgar Allan Poe's well-known essay and selected short stories by Herman Melville, Edward Page Mitchell, and Ambrose Bierce. Chapter

Four is concerned with the proliferation of both interest and success in the production of thinking-machines during the 1940s and 1950s, focusing on Norbert Wiener's work in cybernetics and Alan Turing's universal machines. The Conclusion takes up contemporary debates, suggesting that the new patterns that emerge from these transformations, and the uncertainties about (human) identity that they produce, provoke anxiety precisely insofar as they persist in conceiving "what matters" as a substance rather than as a function.

Copyright © 2009, 2022 Dawn E. Lausa
All rights reserved.

Table of Contents

Acknowledgements vii

Introduction 1

Chapter One/First Iteration: Descartes' Thinking-Machines   15

Chapter Two/Second Iteration: Babbage's Thinking-Machines   58

Chapter Three/Third Iteration: The Automaton Chess-Player and Other Fictional Thinking-Machines   126

Chapter Four/Fourth Iteration: Transformations   184

Conclusion: Identity and Complexity   244

Appendix: Thinking-Machine Timeline   266

Bibliography   268

ACKNOWLEDGEMENTS

These pages are dedicated to John Stewart.
"Your song is all that I am."

This dissertation has taken a circuitous course, and somewhat more than the usual number of years to complete; and so I am perhaps more than the usual amount grateful to those without whom it would not have been completed.

In this light, I am deeply indebted to my committee: Branka Arsic and Gail Hamner, who signed onto the project before knowing much about me or my work; and Don Morton, who knew me but signed on anyway. I am also grateful more than words can express to my director, Gregg Lambert, whose faith, encouragement, gentle prodding, personal and professional ethics, and knowledge of how-things-work have been essential to the completion of this project, and whose dedication to making it happen has gone well beyond the line of duty.

A special thanks also goes to Terri Zollo, whose indomitable good humor and administrative efficiency have rescued me time and again.

No intellectual life develops as a closed system, but as a series of ruptures, collisions, captures, and between-twos; for these I am especially indebted to Doug Hyman, Frank Wilson, Virgil Mann, William C. Hartel, Steven D. Blume, Jim Comas, Dympna Callaghan, Steve Cohan, Bennet Schaber, John Crowley, Michael Martone, Linda Alcoff, Margaret Himley, Peter van Inwagen, and José Benardete.

I am also deeply grateful to Julie Haught, Rick Eckstein, Cathy Burns, and Amy Bates for late-night coffee and afternoon beer; to Don Ruiz, my best writing-buddy and fellow "keeper of the flame"; to Mick Clarke and Gordon Goodykoontz for keeping me sane; to Ken Lause for intellectual rigor; to Mary Lause for "cooler water, higher ground"; to Darren Lausa for growing into one of my best friends; and most especially to Peter Peacock, whose love, sacrifice, support, and patience during the writing process have given me a "reason to rise."

Finally, these pages are also dedicated to my father, Charles A. Lausa, who always encouraged me to take my own lines of flight, and whose image of me was never limited by identity categories like "child" or "girl."

INTRODUCTION

> "[E]ach concept will therefore be considered as the point of coincidence, condensation, or accumulation of its own components. [. . .] neither constants nor variables, but pure and simple *variations*" (Deleuze and Guattari *What is Philosophy?* 19-20).

Thinking-machines
On the face of it, a thinking-machine would appear to be a contradiction in terms. The very nature of a machine seems to signify determinacy, to preclude it from calling upon the kind of indeterminate creativity that characterizes human thought. Discursive formations pointing to the pervasiveness of the association between the machine and unthinking, deterministic behavior are commonplace. Thus, we may defend a small lapse in driving attentiveness by explaining that we were "on automatic," while we describe repeated and thoughtless activities as "mechanical." Without the active engagement of our thoughts, these usages suggest, we humans are mere machines. We are likewise familiar with images that stress the vulnerability of human individuality to the mindless, machine processes that we encounter on a daily basis: Melville's pale girls, like forever blank pages, enslaved to the machinery of the paper mill; Chaplin's Little Tramp who in *Modern Times* is first assaulted by a feeding machine, then fed into the enormous cogwheel machinery which he seems to serve; 1960s youth, responding to a growing sense that they are somehow themselves merely components of a mindless social machine, donning T-shirts which proclaim, "Do not fold, spindle, or

mutilate." In each case, the human is figured as grist for the mill, fodder for the machine's unthinking, unyielding, inevitability.

But, paradoxically, there are also many discursive formations which associate the computer with reason. Thus, we might refer to a particularly intelligent person as a "computer," or describe a coworker who seems lacking in humanity as "calculating." And, despite the many images of unthinking, deterministic mechanism embedded in our language, literature, and popular media, the trope of the thinking-machine is nonetheless also a pervasive one, whether in the guise of the friendly android Data, or the disturbing image of HAL.

Contemporary genealogies of the thinking-machine typically begin with sketchy accounts of ancient Egyptian and Greek oracles, animated statues, and automata designed by gods or men, with various relations to presumed historical accuracy. These include, among a host of other examples, Daedalus' statues, Hero's mechanical birds, the bronze figure Talos patrolling the shores of Crete, and a variety of ingenious oracles equipped with moving parts and echo chambers designed to assist priests in defrauding supplicants. Such genealogies usually nod also toward artificial "men" of non-mechanical contrivance, such as the legendary servant of Rabbi Löw and Mary Shelley's *Modern Prometheus*, an only slightly more scientific version of Löw's golem.[1] The one feature that these artificial creatures seem to share, even if only as a function of lost time and records, is that they are all animated by some mysterious art, breath, or magic that we cannot hope to understand or reproduce.

By the eighth century, however, a very different kind of artificial human image begins to appear, one which is intricately bound up with science, mathematics, and mechanism--that is, with the instruments of reason. It was Arab mathematicians who first integrated moving animal and human figures into their elaborate clockworks and imagined a sort of mathematical thinking-machine (McCorduck 9-11). In Europe, during the thirteenth century, Ramon Lull also invented a thinking-machine. The *Ars Magna* connected reason, logic, formal notation, and mechanism as systematically equivalent and automatic operations for yielding determinate results, directly inspiring Leibniz's calculus as well as his mechanical calculating machine (Cohen 33, 112; McCorduck 11, 26).

Taking a different line of flight from the Arab mathematicians, small clockwork mechanisms that could realistically reproduce human and animal bodily functions for the amusement of onlookers were already well advanced, if not exactly commonplace, by the mid seventeenth century, and automata such as Vaucanson's duck and Jaquet-Droz's "Musical Lady" were widely and famously exhibited throughout England and Europe during the eighteenth and nineteenth centuries. But, by the end of the nineteenth century, the mechanical reproduction of the human body's movements had also been widely exploited by manufacturing interests including, importantly, the weaving industry; and in this real sense, human labor and the human body itself were increasingly supplanted by machines. But, if Descartes was right, these automata which reproduced human bodies and human labor could not yet fully replace the human. Only a machine which could think could do that. And, it is therefore within this milieu that both the automaton chess-player and the calculating engines of Charles Babbage were able to ignite such intense anxiety and foment such popular controversy.

The logic underlying the problematic posed by these very different iterations of the thinking-machine was this: if such mechanical bodies could be made not only to card wool and weave cloth, but also to think, then there would no longer be a need to posit a soul to do that special human labor, and no clear reason why the human should not be replaced by machines altogether. Thus, by the end of the nineteenth century, the trope of the thinking-machine, expressed most iconically as the automaton chess-player and the Difference Engine, had become a staple of literature in Europe and the United States, and a topic of great popular as well as scientific interest. And the question of whether or not a machine could be said to think had taken on a degree of importance that still informs twentieth-century projects in cybernetics, artificial intelligence (AI), and the theory of complex systems (complexity theory).

What I will mean, then, by *thinking-machine* for the purposes of this project is precisely the kind of thing that Descartes imagined when he claimed that such a thing was impossible to imagine; that is, a mechanical device constructed by humans, which is capable of reproducing not merely the bodily motions of the human, but also the (essential) function of human thought. And, it is this image of the thinking-machine, with its genealogical roots in clockwork mechanism, mathematics, logical notation, and the simulation of bodily actions, which condenses and expresses through its multiple iterations the ontological opposition(s) between the human and the machine, the conceptual constellations that identify them, and the material and technological challenges to their separate identities.

Thus, the dissertation does not trace a linear history of these concepts' progressive development over time, nor identify a Kuhnian paradigm shift. And it certainly does not attempt to exhaustively catalogue textual appearances of the thinking-machine. Rather, the object is to carefully sample a much more complex dynamic in which the familiar constellations of concepts that identify the human and the machine are multiply transformed, destabilizing the boundaries between them, and producing multiple and contested new concepts and conceptual constellations which nonetheless never fully replace the old.
In other words, the figure of the thinking-machine in general, and the chess-playing machine in particular, function as sites of contestation, where discourses surrounding the possibilities for machine functionality frame more urgent questions about the nature of the human, ultimately enabling the concepts of the posthuman and complexity to emerge. In short, we care about the ontological status of the machine precisely because it has implications for our own.

Chapter One begins by examining Descartes' concept of the human, and the significance of its identification in opposition to the machine. While Descartes works within (or against) a discourse of scholasticism, and appears to make strong claims for substance dualism, his concepts of the human and of the machine are primarily defended through a discourse of function which tends to undermine his apparent ontological claims. Thus, for Descartes, the human describes a point of intersection and identity among its component concepts which include the functions of speech, reason, will, intelligence, and contingency generally, while the machine is identified with and as the automatic, mindless, and deterministic. However, precisely because they are expressed as functions, the effect of Descartes' multiple iterations of these oppositional concepts is to render the relationships among their components contingent rather than essential, and to multiply the Cartesianisms which emerge both from within and in relation to his own texts. Moreover, in denying the conceivability of the thinking-machine, Descartes has already produced it as an iteration of problems and concepts which will emerge in the wake of Babbage's actual thinking-machines, and as the mechanism through which the concepts of cybernetics, complexity, and the posthuman will emerge in the mid-twentieth century. In this way, Descartes is forced to conceive the very thinking-machine which is strictly inconceivable from within his own ontology.

Chapter Two examines the calculating engines of Charles Babbage as material iterations of the thinking-machine, and the discursive formations surrounding them. Styling himself as an iteration of Descartes' lone and skeptical thinker, Babbage effectively redistributes the function of thought across the boundary between human and machine. Conceiving his calculating machines as an analytic of reason, Babbage sought to embody those deterministic rules which, like those governing the human mind or nature, could develop contingently, without intervention on the part of their creator. That Babbage's calculating engines were widely figured as thinking-machines is evidenced by archival sources including newspapers, popular science writing, and Babbage's own texts. Moreover, while the trope of the thinking-machine becomes commonplace in accounts of Babbage's engines, so too does uncertainty about the boundaries between, and thus the ontological status of, both the human and the machine.

Chapter Three considers this ontological uncertainty as the source of anxiety surrounding Maelzel's automaton chess-player, as evidenced in numerous texts, including Edgar Allan Poe's well-known and derivative essay, and other selected literary iterations of the thinking-machine by Herman Melville, Edward Page Mitchell, and Ambrose Bierce. Within each text, functional capabilities slip unpredictably back and forth across the ontological boundary between human and machine, forcing the conceptual components of both into a multiplicity of new alignments and constellations, destabilizing the clear identity of each. The fact that Babbage's calculating engines actually performed, or could have performed, contingent thought-like functions while Maelzel's chess-player was a hoax did not prevent the chess-player from performing a similar function with regard to the transformation of the concepts of the human and the machine. Indeed, as a material iteration of the thinking-machine, Maelzel's chess-player arguably presented the more affective image. Moreover, the image of Babbage's thinking-machines is so commonly invoked by texts responding to Maelzel's fraudulent automaton that objections to the chess-player must be read, at least in part, as anxiety about the material undeniability of Babbage's engines. In both cases, the trope of the thinking-machine effects a transformation through which the category of the human loses its ontological status, as identifying human functions are usurped by machines, which threaten to master their human creators.

Chapter Four is concerned with the enormous proliferation of both interest and success in the production of thinking-machines during the 1940s and 1950s. Catalyzed by military efforts during World War II, a variety of separate projects on machine intelligence for purposes as diverse as code-breaking and targeting-control converged around problems of machine language, symbolic logic, and communication, and the related (mathematical) problem of providing a machine with determinate instructions for engaging in contingent behaviors. Interestingly, but not surprisingly, many of the researchers in this field also designed and built chess-playing computers alongside their work on machine languages and programming.[2]

Among the most prominent and successful researchers in these areas were Norbert Wiener and Alan Turing. In many ways, they are also among the most interesting, in part because, more than most others, they spent a great deal of time thinking and writing about the philosophical and social implications of their work. Moreover, while in some technical aspects their research took very different approaches, their theoretical thinking also converged in many ways.[3] In 1949 and 1953 respectively, two important books appeared with the goal of summing up the progress of the previous decade's work on computers and related technologies, and of exploring what had become a fashionable question: Can machines think? In answer to this problem, Edmund Berkley in *Giant Brains* wrote that, a "machine can handle information; it can calculate, conclude, and choose; it can perform reasonable operations with information. A machine, therefore, can think" (Berkley *Giant Brains* 5). B. V. Bowden, to the contrary, looking at the very same machines and the very same machine capabilities, insisted in *Faster than Thought*, that, "we do not claim that machines think for themselves. That is precisely what they cannot do" (Bowden *Faster than Thought* 29). Both books, however, devoted a considerable effort to discussing what might be meant by words like "machine" and "think."

But, as we have said, these concepts are intimately bound up with a Cartesianism that identifies the concept of the human with the component concepts of thought, reason, will, and contingency in contrast to the deterministic, unthinking machine. In other words, the question, Can a machine think? is grounded in a conceptual schema where the function of thought is coextensive with the concept of the human, at the same time that the material technologies motivating the question have transformed these conceptual relations beyond recognition. As a result of this transformation, the function of thought is distributed over such a broad ontological territory, that the separate categories of human and machine--and the questions they ground--no longer seem sensical.

It is within this context that both Wiener and Turing offered answers to the question, Can machines think?, that might also be appropriately characterized as transformations. That is, rather than accepting the terms of the question, or even attempting simple redefinitions, both Wiener and Turing engaged in multiple manipulations of language and concepts that undermined the ontological stability of the human and the machine, while the posthuman concepts of cybernetics and artificial intelligence began to emerge. Moreover, through these transformations, lingering anxieties about the mastery of the human by the machine were also transformed, as anxieties over uncertain ontological categories became anxieties about the functional dominance of determinism in increasingly rigid systems.

The Conclusion addresses persistent deep anxieties concerning the loss of the human and of identity in general, despite the transformations effected by cybernetics and AI. For Descartes, working within, and against, the discursive constraints of scholasticism, the human could (only) be conceived and identified as the substance of contingency, as opposed to the determinacy of mechanism; in this way, both contingency and human identity *mattered*. But, already within Descartes, the concept of contingency, expressed as speech, will, reason, soul, thus folded into its own material actualizations, became the *function* of contingency. And this function, once sundered from substance, could be infinitely iterated both within and beyond the human. Babbage's calculating engines and Maelzel's chess-player provide examples of such iterations, redistributing the component concepts of the human and the machine across categories, and into multiple and contingent constellations, where the boundaries between the human and the machine become uncertain. For nineteenth-century American writers like Poe, Melville, Mitchell, and Bierce, as well as for contemporary theorists like Hayles and Haraway, the patterns that emerge from these new constellations, and the uncertainties about (human) identity that they produce, provoke anxiety precisely insofar as they persist in conceiving contingency as a substance rather than as a function. For it is only within this logic that the loss of human identity appears to threaten the contingent, creative processes of becoming that really matter.

CHAPTER ONE/ FIRST ITERATION: DESCARTES' THINKING-MACHINES

> Vizzini: "Inconceivable!"
> Inigo Montoya: "[. . .] that word. I do not think it means what you think it means" (Goldman *The Princess Bride*).

Introduction

"Descartes"--the signature is synonymous with metaphysical dualism, the separate substances of body and soul, the basis for a view of special human identity that transcends the material, and so distinguishes us categorically from both beasts and automata—the view that Gilbert Ryle famously critiqued as "the dogma of the Ghost in the Machine" (Ryle 15). Lillie Alanen calls this caricature of Cartesian dualism the "myth of the Cartesian myth," writing that, "[i]n spite of its popularity in Anglo-American philosophy of mind, this Rylean version of Cartesian dualism has not much in common with the view Descartes actually held" (45). Alanen goes on to point out that Ryle's sketch of Cartesian dualism is indeed an accurate portrayal of the metaphysics of the "Cartesianism" that has persisted despite Descartes.

"I am not merely present in my body as a sailor is present in a ship," Descartes wrote, "but am very closely joined and, as it were, intermingled with it" (*Meditations* 56). While the passage seems to imagine the "I" as something like an emergent property of the body rather than as a separate substance, it is the image of the sailor in the ship, after all, that most clearly describes the Cartesian myth/legend, and which persists as the metaphysical ground for the human.

In order to avoid confusion with Ryle, Baker and Morris have called this the "Cartesian Legend," "a second order myth, a myth about a myth" (1). Schmaltz describes instead, "a collection of systems, deriving from Descartes" (xi-xii). Thus, the most recent projects on Descartes aim, on the one hand, to revisit Descartes' historical body of work in order to recover the authentic Descartes from Cartesianism, and on the other, to trace the multiplicity of Cartesianisms within their cultural and intellectual milieus.

Perhaps more interesting than any particular reading of Descartes' metaphysics, however, is the way in which his ontological schema is discursively framed in terms of function. Thus, whatever Descartes may have had in mind, and whichever reception one considers, the signature most associated with the ontological separation of mind and body (read as contingency and determinism), instead works at some level toward conceiving thought as an emergent property of the body (that is, as a complex system emerging from the play between contingency and determinism). Thus, as Rosenfield remarks, "The more one showed the highly organized perfection of the bodily machine, the easier it became to take the jump and proclaim that psychological processes in man as well as beast consist only of physiological activity" (22).

This, of course, is one of the most notorious problems with Descartes' dualism, and is already raised by Hobbes in his Objections to the Meditations.[4] This is also the weakness which will be exploited to dramatic effect by La Mettrie, whose materialism represents, in Cartesian terms, the very end of the human. Not surprisingly, then, it is also with reference to Descartes that both the substantial separateness of the thinking thing and the pervasive infusion of soul within all matter are defended.[5] In other words, even if we distinguish between Descartes and Cartesianism, there is no single or simple Cartesianism to which we can refer.

Thus, despite the unifying power of the philosopher's "conceptual persona," there exist a multiplicity of actual Descartes (as well as a virtual infinity of them). And this multiplicity exists not only within the texts historically attributed to and signed by Descartes himself (where it does indeed exist), but also in the constellation of texts in which the name Descartes is invoked for one purpose or another. What all of these anti-/Cartesianisms share in common, however, is the identification of thought as the last uniquely human function to be usurped by mechanism. As such, for all these anti-/Cartesianisms, it is thought (and therefore contingency) upon which the ontological status of the human hinges, and it is around this problematic that efforts to defend or breech the domain of thought must turn.

Moreover, while at one level, both mechanism and vitalism iterate variations of substance dualism, to the extent that these iterations of substance rely upon functional arguments, they are already participating in the emergent concepts of cybernetics and the posthuman, for which the important questions are not about the material substance of being, but about the dynamic functional relations between contingency and determinism.[6] Thus, Descartes, functioning as the fulcrum for the mechanist/vitalist debate, sets up thought as the last essentially human function, thereby conceiving (despite himself) the thinking-machine.

It is not my purpose, therefore, to identify (or to critique) Descartes' authentic concept of the human, or of thought, but to consider the ways in which Descartes' exploration of the human put into play the conceptual components that enabled his philosophical heirs to conceive the constellation of concepts that encompass cybernetics, complex-systems, and the posthuman. But, "every concept has a *history*, even though this history zigzags, though it passes, if need be, through other problems or onto different planes," and "every concept relates back to other concepts, not only in its history but in its becoming or its present connections" (Deleuze and Guattari *What is Philosophy?* 18-19). In this sense, "Descartes" is as much a product of cybernetics and the posthuman as they are of Descartes.

I admit also, therefore, that I am less concerned at any given moment to distinguish between Descartes and Cartesianism, or to identify the real Descartes within his own texts, than I am with considering the discursive fields--the shifting sets of concepts and their oppositions--that are put into play by Descartes around the ontological problem of the thinking-human, and whose multiple iterations (from the beginning, but also more self-consciously in response to the calculating machines of Charles Babbage) include the figure of the thinking-machine.

Cogito Ergo Humanus
> "it is not conceivable that such a machine should produce different arrangements of words so as to give an appropriately meaningful answer to whatever is said in its presence" (Descartes *Discourse* 140).

It is inconceivable, Descartes argues, that an automaton could be made to think--not simply impossible, but *inconceivable* (*Discourse* 140). This is because Descartes' concept of thought entails the concept of contingency (evidenced by speech), and his concept of mechanism entails the concept of determinacy. Indeed, each of these concepts functions as an iteration within the opposing and mutually dependent systems of concepts describing the myth of Cartesian mind/body dualism, where contingency reliably identifies the human, thought, soul, reason, speech and will,[7] while the animal, the material, mechanism, automata, and the body are identified by determinacy. Given Descartes' concepts of thought and mechanism, then, the inconceivability of a thinking machine is nothing other than the *a priori* observation that (just as it is impossible to conceive of a square circle) it is impossible to conceive of deterministic contingency. That is, the concepts are mutually exclusive. Descartes' denial of the possibility of machine intelligence, in other words, is not a statement about what science might or might not one day achieve; rather, it is literally a statement about what it is (not) possible to imagine.

But, this claim is also dependent upon Descartes' specific concepts of thought and mechanism (mind and body) as separate substances.[8] That is, as long as Descartes' associations of contingency with thought, and determinism with mechanism, are given metaphysical status, there is simply no way for mechanism to ever become mind, or to be conceived as such. From a logical point of view, it is strictly inconceivable.

The problem for Descartes (and for Cartesianism), however, is that the metaphysical status of Descartes' dualism is far from clear, even in his own writing, and the opposing systems of concepts which it describes have a tendency, with every iteration, to shift, intermingle, and collapse into one another in unexpected ways. That is, it is through the discursive multiplicity of anti/Cartesianisms that the concepts of its central metaphysical oppositions are folded and refolded like layers of dough,[9] shifting and transforming the component concepts of the human and the machine--contingency and determinism, mechanism and vitalism, thinking and feeling, reason and passion, being and function, body and soul--into new conceptual systems and alignments.

Central to this discursive multiplicity are four distinct but related premises established by Descartes: (1) that both animal and human bodies are the functional and (therefore) ontological equivalents of machines; (2) that machines and body-machines function deterministically according to fixed laws and design, producing fixed outcomes; (3) that thought (variously termed mind, intellect, will, soul, or reason) is the function of indeterminacy and contingency; and (4) that, for all of these reasons, the *ontological* essence of human identity —that is, its distinction from both animal and machine identities--is precisely the *function* of thought, or reason.[10] It is, however, just this slippage between substance and function that opens Descartes and Cartesianism to the emergent concepts of cybernetics and the posthuman.[11]

Body-Machines
For Descartes, the human body is the functional and ontological equivalent of the machine. As such, it is not in the body that a separate human identity can be located. In a sense, as far as bodies are concerned, there is no such *thing* as a human being. The category is ontologically empty. This, as both Gaukroger and Rosenfield suggest, allows Descartes to identify the essentially human as something separate from the mechanistic universe of Newton and Galileo, while making the body a legitimate object of scientific inquiry (Gaukroger 11-14, passim; Rosenfield 23). Thus, Descartes attributes all non-rational[12] human activity to the mechanistic/animalistic body, collapsing all bodies into the mechanistic universe: "the bodies of animals contain all the organs which an automaton needs if it is to imitate those of our actions which are common to us and beasts" (Descartes *Correspondence* 149). Importantly, Descartes' language here also accomplishes a discursive shift from substance to function.[13] Human beings uniquely *are* the separate substance of thought precisely because human bodies, animal bodies, and automatons do not think. It is the *"action"* of thought, then, which defines the human, and the lack of that *action* which renders all bodies equivalently mechanistic.[14]

But, while Descartes' identification of machine and human bodies will become significant for later discursive practices surrounding questions of machine intelligence, his immediate concern is to refute claims that animals have souls, and to support an essential difference of kind between human beings and animals.[15] He quite takes for granted that no reasonable person could ever mistake a machine for a human being, though he goes to some lengths in Part Five of the *Discourse* to explain why he thinks such a confusion would be inconceivable. Throughout the *Discourse* and *Meditations*, as well as in his letters and other extant texts, Descartes uses the trope of the automaton to describe both human and animal bodies, and to thereby insist upon the separate functioning of the body from the soul--that is, "that the soul [is] a substance really distinct from the body" (Descartes *Correspondence* 100).

Descartes logically and discursively equates the body with "anatomy and mechanics" (*Description of the Human Body* 314), using constructions such as "this mechanism of our body" (*Correspondence* 334), "mechanism of the human body" (*Meditations* 58), and "the whole mechanical structure of limbs" (*Meditations* 17). Beyond the pervasive use of these incidental phrases, several of Descartes' texts focus primarily on describing "the entire bodily machine" (*Description of the Human Body* 315), most notably *Treatise on Man* and *Description of the Human Body and All its Functions*. Both deal specifically with "bodily movements," or, negatively defined, those functions which "do not involve any thought" (*Description of the Human Body* 314-315).

Drawing largely on Harvey, as well as his own animal dissections,[16] for both texts, Descartes describes the heart as "the great spring [. . .] responsible for all the movements occurring in the machine," the veins as "pipes which conduct the blood from all the parts of the body towards the heart," where it is heated so that "the parts of the blood that are most agitated and lively" are carried by the arteries, "yet another set of pipes" to the brain.[17] From there, these "animal spirits . . . inflate the muscles in various ways and thus impart movement to all the parts of the body" (*Description of the Human Body* 316). Striking for its strictly mechanistic understanding of the circulatory system, Descartes' description of the living human body does not merely metaphorically liken the body to an automaton, but minutely describes it *functionally* as such. That is, for Descartes, it is only a lack of skill that prevents us from producing machines as intricate as those of the human body, since both rely upon the same mechanical principles, and are different by degree, not by kind.[18]

Also interesting is Descartes' mechanistic account of the interface between the body and the outside world, with which it is one substance. In a detailed analysis of the pain one feels when standing too close to a fire, Descartes writes that

> the tiny parts of this fire (which, as you know, move about very rapidly) have the power also to move the area of the skin which they touch. In this way they pull the tiny [nerve] fibre [. . .] attached to it, and simultaneously open the entrance to the pore. [. . .] the animal spirits enter [. . .] and are carried through it—some to muscles which serve to pull the foot away from the fire, some to muscles which turn the eyes and head to look at it (*Treatise on Man* 101-102).

This passage is echoed in the *Meditations*, where Descartes writes,

> when I feel a pain in my foot, physiology tells me that this happens by means of nerves distributed throughout the foot, and that these nerves are like cords which go from the foot right up to the brain. When the nerves are pulled in the foot, they in turn pull on inner parts of the brain to which they are attached, and produce a certain motion in them (60).

Again, Descartes describes an entirely mechanical reaction of the human body-machine to danger and pain, independent of any intervention of soul (that is, of thought, reason, or will). This concept of the mechanical contains the concept of determinacy, and is discursively linked to it, such that the absence of soul or reason is precisely the absence of contingent function.

Thus, implicit in Descartes' understanding of the body-machine is its clock-like automatic functioning. The clock, of course, with its precisely measured performance, was and is a powerful trope signifying both predictable determinacy and the efficient (and dehumanizing) mechanical regulation of human activity.[19] For Descartes, too, the clock signified the ultimate knowability of the mechanical universe. Furthermore, if automata are "the forgotten ancestors of all modern technology" (Standage 2), then surely the highly complicated municipal and cathedral clocks of medieval Europe--which served as the initial inspiration for the smaller, privately owned automata produced for wealthy patrons, often by the same clock-makers--are even more fundamentally ancestral to subsequent thinking-machines. Like the automaton in general, then, the clock can be understood both discursively and in terms of its real material impact on social organization and subjectivities (McLuhan *Understanding Media* 145-156).[20]

It is not surprising, then, that Descartes, who was "influenced by the automata that were proliferating throughout Europe in chateau gardens and municipal clock towers" (McCorduck 39),[21] should invoke the trope of the clock —that most perfect automaton--specifically and repeatedly.

For instance, in his 23 November 1646 letter to the Marquess of Newcastle, Descartes uses the clock in an argument refuting the proposition that animals have souls. Though tautological, his argument is revealing. "I know that animals do many things better than we do . . . it can even be used to prove that they act naturally and mechanically, like a clock which tells the time better than our judgment does. Doubtless when the swallows come in spring, they operate like clocks" (*Correspondence* 304). It is the precision of the swallows' return, their ability to find their way at exactly the right time, without reference to map or compass that interests Descartes. This passage, through the trope of the clock, then, makes a discursive link between mechanism and determinism, implicating the swallows in the process, so that the concept of the former essentially includes the latter.[22]

Similarly, in describing the mechanical movements of the human body as a function of the animal spirits,[23] Descartes draws this detailed metaphor:
> one may compare the nerves of the machine I am describing with the pipes in the works of these fountains, the muscles and tendons with the various devices and springs which serve to set them in motion, its animal spirits with the water that drives them, the heart with the source of the water, and the cavities of the brain with the storage tanks. Moreover, breathing and other such activities which are normal and natural to this machine, and which depend on the flow of the spirits, are like the movements of a clock or mill (*Treatise on Man* 100-101).[24]

Again, Descartes stresses the *deterministic* nature of the human body through the image of the *automatic* clockwork.

Later, in the *Meditations*, Descartes compares "a sick man and a badly-made clock, and the idea of a healthy man and a well-made clock" (59), and describes
> the body of a man as a kind of machine equipped with and made up of bones, nerves, muscles, veins, blood and skin in such a way that, even if there were no mind in it, it would still perform all the same movements as it now does in those cases where movement is not under the control of the will, or consequently, of the mind but occurs merely as a result of the disposition of the organs" (58).[25]

Descartes' language here is important: the human body is not *like* a machine; it *is* a "*kind of machine*," and so really incidental to the quality of being human. In other words, the deterministic functioning of the body is opposed to the contingent functioning of the mind, and it is in that contingency of thought, will, soul--not in the automatic body--that human identity is located.[26]

Descartes also stresses the mechanistic nature of the body (as opposed to the contingent thinking of the soul) when he writes that "when all the body's organs are appropriately disposed for some movement, the body has no need of the soul in order to produce that movement" (*Description of the Human Body* 315). He goes on, he hopes, to

> give such a full account of the entire bodily machine that we will have no more reason to think that it is our soul which produces in it the movements which we know by experience are not controlled by our will than we have reason to think that there is a soul in a clock which makes it tell the time (315).

And, again, following his discussion of the mechanisms of the heart, Descartes writes that "[t]his movement follows just as necessarily as the movement of a clock follows from the force, position, and shape of its counter-weights and wheels" (*Discourse on the Method* 136). If there is any doubt, then, that Descartes takes the movement of clockworks to be deterministic, his characterization of their movements as a "necessary" outcome of their arrangement should dispel it.

In each case, for Descartes, the clock signifies the essence of the machine in that its "movements" are absolutely determined by its design, which in turn determines "the arrangement of its counter-weights and wheels" (Descartes *Treatise on Man* 108). Furthermore, Descartes' use of the automaton clearly also signifies the deterministic nature of mechanism that Descartes identifies with the body and extension, and this identification of the automaton with the clock is perfectly consistent with the technological development of both.

That is, Descartes' concepts of body and machine both include the component concept of determinism, and the two condense around a zone of indiscernibility. Thus, for Descartes, the clock also signifies the deterministic nature of the body, whose "functions follow from the mere arrangement of the machine's organs" (*Treatise on Man* 108). Therefore, it is not by their bodies that human beings can be distinguished essentially from animals, for both animal and human bodies share both the functional and ontological status of machines. And, these body-machines function deterministically according to their design, like clockwork.

Thus, Descartes' claims about the *ontological* distinction between the separate substances of body and the soul (thought), and their opposition in relation to contingency, are framed discursively in terms of *function*. And, it is the difference between deterministic and contingent functioning that supports his ontological distinctions between human and animal, human and machine, and human mind and human body. For each of these oppositional binaries, of course, it is human thought, conceived as reason, mind, will—that is, unified subjectivity or consciousness--which serves as the privileged term and, indeed, as the human itself.[27] But, once the element of function is introduced into the concept of identity, these binary pairs become decoupled, and are freed to condense in new configurations around new concepts.

The Rational Soul
This can be seen through a closer examination of Descartes' discourse surrounding the trope of the soul. Though Descartes often explicitly writes that he understands the soul to be a body-independent substance,[28] (that is, presumably, an ontological entity), it is also the case that he often seems to use the soul as a trope for the various *functions* of the conscious mind-- that is, for the *contingent* functions of thought not identified with the *deterministically* functioning body. On this reading, the thinking-soul and the body-machine operate as tropes signifying contingent and deterministic functions respectively, and, again, the discursive ground for human identity becomes functional rather than ontological.[29]

In a letter to Regius dated may 1641, for example, Descartes uses typical language in a refutation of the traditional "three-fold soul." Here Descartes insists that "There is only one *soul* in human beings, the *rational soul*; for no actions can be reckoned human unless they depend on reason" (*Correspondence* 182). Later in the same letter, Descartes objects to calling the vegetative and sensory powers of the body "souls" because "since the *mind, or rational soul*, is distinct from the body . . . it *alone* is called the *soul*" (*Correspondence* 182). Here, as elsewhere in Descartes' texts, the soul is discursively framed as reason—that is the ontological *substance of soul* is discursively framed as the *function of contingency*.[30]

Of course, in the Second Meditation, Descartes writes, "I exist. [. . .] as long as I am thinking. [. . .] I am a mind, or intelligence, or reason. [. . .] a thinking thing" (*Meditations on First Philosophy* 18). Here the haecceity of the "I" seems to rely upon the activity or function of thinking, to name thought itself, so that the "thinking" in "I am thinking" can be understood both as a present participle and as a gerund. While at this point in the *Meditations* Descartes' conclusions must be taken as provisional, with care "to grasp the proper order of [the] arguments and the connections between them" (*Meditations on First Philosophy* 8) this language is consistent with language in other of Descartes' texts, as well as with the concluding chapter of the *Meditations* itself, where he reiterates that "my essence consists solely in the fact that I am a thinking thing" (54).

Even more explicitly, in his Fifth Set of Replies to the *Meditations*, Descartes writes, "I consider the mind not as a part of the soul, but as the thinking soul in its entirety" (*Meditations* 246). Thus, despite a parallel language describing the human as the incorporeal substance of soul, Descartes often seems to conceive thought as an emergent property, a pure function. But, as we have said, this has the effect of disentangling thought's ontological link to the human, and thereby opening a line of flight toward the thinking-machine.

In Part Four of the *Discourse*, in what serves as a summary of the text which will be *Meditations on First Philosophy*, Descartes describes himself–the "I" which doubts—as "a substance whose whole essence or nature is simply to think," and this "I" as "the soul by which I am what I am" (127). In Part One of the *Discourse*, Descartes describes the "mind" or "reason" as "the only thing that makes us men and distinguishes us from the beasts" (112). Thus, given Descartes' belief in the functional and ontological equivalences of animals, the human body, and automata, this amounts to saying that the human is precisely the non-mechanical as well as the non-animal.[31] That is the human soul—the human being--*is* the non-corporeal function of thought.

Thus, because the identity of the human being hinges on the faculty of thought, an important underlying implication of Descartes' insistence "that the soul [is] a substance really distinct from the body" (*Correspondence* 101) and "by nature entirely distinct from s the body" (*Correspondence* 163), is that "no actions can be reckoned human unless they depend on reason" (*Correspondence* 182).[32] This does not amount merely to the tautology that neither animals nor machines can truly think because they are not human, but is rather intended as a conclusion derived from that premise.

But, the association of thinking and reason with human identity creates a discursive field in which the trope of the thinking-machine as it emerges through Charles Babbage's calculating engines and Kempelen's automaton chess-player will threaten the very concept of the human. In short, Descartes has identified thought as the last human function not identified as mechanism.

Speaking-Machines
It is important, as we have seen, to understand that for Descartes there are relations of identity between soul, mind, reason, thought, and human identity (both in the sense of the individual person and of the species), and that these are identified with contingency. Thus, in answer to the question of whether animals might think, Descartes writes that "if they thought as we do, they would have an immortal soul like us" (*Correspondence* 304). Again, for Descartes, this is an *a priori* argument. That is, where animal and human bodies function according to the same laws as do machines, what distinguishes

humans from animals, and the human being from the human body, is the function of thought. Thus, again, while Descartes repeatedly insists that the soul is an immortal and separate substance that survives the death of the body, he tends to think of the soul in terms of *function*—and specifically the function of *thought* (the nature of which is contingency).

Significantly, speech functions for Descartes as the revelation of thought, and speech itself is therefore explicitly described in terms of contingency. Thus, one might always be able to distinguish between a human being and an automaton, since "automatons [would] never answer in word or sign, except by chance, to questions put to them" (Descartes *Correspondence* 99).

Similarly, in Part Five of the *Discourse*, Descartes summarizes a section from his unpublished *Treatise on Man*, using the automaton as a trope to illustrate the essential differences between humans and animals:

> I made special efforts to show that if any such machines had the organs and outward shape of a monkey or of some other animal that lacks reason, we should have no means of knowing that they did not posses entirely the same nature of these animals; whereas if any such machines bore a resemblance to our bodies and imitated our actions as closely as possible for all practical purposes, we should still have two very certain means of recognizing that they were not real men" (139-140).[33]

Again, for Descartes, the differences between human, animal, and machine bodies are not essential differences of kind. What distinguishes human identity is the contingent function of thought. And, tautologically, this contingency cannot be designed into the deterministic arrangement of parts that characterizes automata or body-machines. Thus, Descartes' two "certain" means of identifying the human, speech and contingency, are themselves the very "actions" that he has already defined as uniquely human insofar as he regards them as evidence of thought; and it becomes clear once again that he is arguing tautologically (or asserting rather than arguing at all) when he declares it "not conceivable that such a machine should produce different arrangements of words so as to give an appropriately meaningful answer to whatever is said in its presence" (*Discourse on Method* 140).[34] The sounds of speech might be reproduced by mechanism and made to respond to specific actions, he argues, but it is inconceivable that machines might reproduce the *contingent* nature of real speech.[35]

Descartes' second "certain" telltale regarding the difference between a human being and a machine turns out to explain the first, and to underscore the functional association that Descartes wants to make between speech and thought:

> Secondly, even though such machines might do some things as well as we do them, or perhaps even better, they would inevitably fail in others, which would reveal that they were acting not through understanding but only through the disposition of their organs. For whereas reason is a universal instrument which can be used in all kinds of situations, these organs need some particular disposition for each action; hence it is for all practical purposes impossible for a machine to have enough different organs to make it act in all the contingencies of life in the way in which our reason makes us act" (*Discourse* 140).[36]

Neither beasts nor machines, Descartes continues, are capable of "arranging various words together and forming an utterance from them in order to make their thoughts understood" because "they have no reason at all" (140).[37] That is, as mechanism, they cannot function contingently.

Thus, Descartes identifies speech as the outward sign of hidden thought, a capacity which he believes animals lack. That is, he argues that the absence of true speech among animals is not a failure of the body's capacity to produce the appropriate sounds, but a consequence of an absence of thought to direct them:

> This does not happen because they [animals] lack the

necessary organs, for we see that magpies and parrots can utter words as we do, and yet they cannot speak as we do; that is, they cannot show that they are thinking what they are saying" (Descartes *Discourse* 140).

And, again, Descartes writes to the Marquess of Newcastle in 1646 that, "the reason why animals do not speak as we do is not that they lack the organs but that they have no thoughts" (*Correspondence* 303).

Animals do not speak, then, because animals do not think. Again, Descartes expresses his ontological claims in terms of function. On Descartes' own logic, if animals—mere automata--*were* to speak, this might indicate a capacity for thought; surely, then, a mechanical automaton that can "speak as we do"—that is, contingently and with meaning--is precisely a thinking-machine. This, in fact, is precisely the argument laid out by Alan Turing in 1950. For Descartes--and Turing-- the surest test of intelligence is speech, and this is precisely because speech is *not* automatic, but rather involves contingent play within (more or less) determined rules. Echoing Descartes precisely, Alanen identifies "free choice of the will" as the underlying reason "that the human person counts as a rational agent capable of applying rules intelligently as opposed to an automaton that follows them mechanically" (99).

But, given Descartes' synonymous usage of "will," "reason," "thought," and contingency as identifying human functions, the argument is simply tautological. What the Descartes/Turing test suggests is that a machine which *could* pass the test of "natural speech," or of contingent play in general, could no longer be considered an automaton--that is, its functioning would no longer be strictly automatic or deterministic. What Turing suggests is that, unless we assume from the outset that only humans can think, then, given evidence of a machine that functions as we would expect a thinking-human to function, we have no reason to deny that it is a thinking-machine (Turing "Computing Machinery and Intelligence" *passim*).[38]

For Descartes, then, an important central issue which links and justifies his various intellectual projects is this location of human identity in the soul (that is, mind, reason, thinking, will), and, conversely, the soul's absolute humanity, so that "mind [. . .] is the only thing that makes us men" (Descartes *Discourse* 112). It is around this issue that Descartes develops his concepts of mind and body as separate substances, which allows him to defend the eternal nature of the soul—that is, "that it is by nature entirely distinct from the body, and consequently it is not bound by nature to die with it" (Descartes *Correspondence* 163)—and thereby to help justify his scientific and intellectual pursuits at a time when other natural philosophers are being condemned or imprisoned by the Church.[39] In an attempt to make his argument transparent, Descartes uses the trope of the automaton, already a topic of wide public interest, particularly among "men of reason."

Importantly, it is the *inconceivability* of a machine functioning contingently, as evidenced through speech that reveals the function of thought, that underpins the ontological distinction for Descartes. And, it is just this issue of determinism and contingency that becomes central to the discourses surrounding the trope of the thinking-machine throughout the nineteenth and early twentieth centuries, and which eventually informs the emergent concepts of cybernetics and the posthuman. Indeed, it will be Babbage's ability not only to imagine, but in a limited way to materially conceive, machines capable of contingency that will force shifts in the discourses surrounding both the human and the machine, and the emergence of the new concepts of cybernetics, complexity, and the posthuman.

La Mettrie's Thinking-Machines

Of course, Descartes' unintended thinking-machines are precisely La Mettrie's most notorious invention. La Mettrie applies Descartes' functional criteria for ontological categories, and draws upon contemporary research into the "motive principle" of the body's smallest parts, in order to show that *all* animal and human function can be accounted for by the specific organization of matter (La Mettrie *Machine Man* 26ff). Indeed, using language that echoes Descartes, La Mettrie insists, "the soul's faculties depend so much on the specific organisation of the brain and of the whole body that they are clearly nothing but that very organisation," and the "soul is only a vain term [. . .] which a good mind should use only to refer to the part of us that thinks" (*Machine Man* 26).[40]

Thus, for La Mettrie, as for Descartes, the body (animal or human) is a clockwork mechanism. But, whereas Descartes' body-machine relies upon the soul for its animation, La Mettrie's body-machine is "a machine which winds itself up" (*Machine Man* 7). In both cases, human bodies "are basically only animals and vertically crawling machines" (*Machine Man* 35), but for La Mettrie, the soul is merely the *name* for "the part of us that thinks" (*Machine Man* 26). In other words, once the body's material functions have been elaborated, there is no residual function to suggest the need for a soul. The soul, in La Mettrie's analysis (which he takes pains to ground in empirical, or *a posteriori*, observation), is redundant, and it is a simple application of Occam's razor that eliminates it from his ontology.[41]

It is useless, therefore, to say that La Mettrie argues with or against Descartes, that he is Cartesian or anti-Cartesian. Rather, he takes up the play, fracturing, multiplying, and rearranging the concepts that Descartes has put into play, so that the thinking-human is a thinking-body is a thinking-machine.

Repeatedly, La Mettrie tells us that "it is a folly to waste one's time trying to discover its mechanism" (*Machine Man* 33), and it appears that La Mettrie's primary purpose in *Machine Man* is "not to develop a scientific explanation of the workings of the human 'machine', but to use any evidence to deny the need for a soul" (Thompson xix). The latter argument (that thinking and mind are functions of the body-machine) is merely a corollary to the first (that the human being is a machine); but the distinction regarding La Mettrie's aim is important, since once human mind can be accounted for by a specific organization of matter without recourse to a separate soul--that is once the human body-machine can also be understood as a thinking-machine--it becomes possible to say that any machine designed for the function would be capable of thought.

La Mettrie also has much to say regarding the possibility of speech in animals. Like Descartes, he considers the function of contingent speech[42] to be a reflection of thought,[43] and distinguishes between this and the mere repetition of phrases and tunes, citing "a parrot who replied pertinently and had learnt, like us, to conduct a sort of coherent conversation" (*Machine Man* 12). Significantly, it is "the similarity of the ape's structure and functions" to our own that encourages La Mettrie to suggest that "if this animal were perfectly trained, we would succeed in teaching him to utter sounds and consequently to learn a language" (*Machine Man* 12).

Thus, for La Mettrie, *function*--the direct effect of organization--*is*, unproblematically, ontology. Because, "[f]rom animals to man there is no abrupt transition" in function, there is also no clear ontological boundary between animal and human (*Machine Man* 13). But, if that is the case, then it is also the case that there is no clear ontological boundary between the human and the machine, and no particular reason to believe that a machine could not, at least in principle, be made both to speak and to think.

There is a story told of Descartes and a life-sized mechanical replica of his daughter, Francine, which, while undoubtedly false, nonetheless illustrates something important about the impact of Descartes' automatism for later generations.[44] The story always begins with Descartes traveling at sea with the automaton. In some versions, the two are frequently seen together aboard ship, appearing inseparable; in others, Francine is never seen in public. Commonly, the automaton sleeps inside a wooden trunk at the foot of Descartes' bed. And, invariably, the crew--aroused by some uncanny suspicion or curiosity, often at night--discover her there, hurling her, in their supreme horror, overboard.[45]

While Wood, citing Descartes' earlier interest in constructing automata, postulates that the story might be a true account of a grieving father's attempt to recover his dead daughter, it is most certainly not that (Wood 4-5). For, the construction of an automaton as a replacement for a human being is precisely what Descartes found inconceivable-- and his entire metaphysics depends upon that. The story points, rather, as both Gaukroger and Rosenfield suggest, to the anxieties surrounding anti-/Cartesianism as it developed along mechanist, or materialist, lines, threatening religious doctrine, and leaving little need for either God or the spirituality of the human soul (Gaukroger 1-2; Rosenfield 294). In other words, while the anxiety over the machine who steals the human soul (and, therefore, human identity) emerges fully and overtly only through writers like La Mettrie, who follow in the wake of Descartes, it is with Descartes that the story locates the origin of its anxiety, lurking, hidden in the dark, concealed among the baggage of Descartes' dualism--the site at which the mirror images of the machine-man and the thinking-machine are first conceived.

CHAPTER TWO/ SECOND ITERATION: BABBAGE'S THINKING-MACHINES

What shall we think of the calculating machine of Mr. Babbage? (Poe "Maelzel's Chess-Player" 348).

"'[I]s not a closed system the essence of the mechanical, the unthinking? And is not an open system the very definition of the organic, of life and thought?

If we envision the entire System of Mathematics as a great Engine for proving theorems, then we must say [. . .] that such an Engine *lives*'" (Ada Byron Lovelace fictionalized in Gibson and Sterling *The Difference Engine* 421).

Introduction
Poe's question, of course, is more complex than it appears. Articulated from within the discursive formations of a reductive Cartesianism, it presumes that the subjective "we" is human, and that, therefore, it is only the human "we" that can function as thinking subjects. The form of the question structures the answer: humans and machines are essentially opposed

because humans are capable of the function of thought, while machines are not. Thought, for Poe, as his essay on the chess-player makes clear,[46] is defined by contingency; and Babbage's machines, in Poe's conceptual schema, do not think precisely because they function only deterministically. Thus, the baker's transformation can be used to describe the event of Babbage's calculating engines, and the destabilizations and transformations in the concepts of the human and the machine that they provoke; or, indeed, to map the movement between what Poe and what Gibson and Sterling think of Mr. Babbage's calculating machine.

Importantly, for Charles Babbage, however, the laws of thought were also expressed as the laws of mathematics, whether through the Leibnizian notation for calculus, through the system of punch cards for the Analytical Engine, or through the mechanical arrangement of cogs and wheels that comprised the calculating engines themselves.[47] Each of these systems, in addition to its practical efficacy, served also for Babbage as an *analytic of thought*.

Poe's question, then, iterated within the problematic presented by Babbage's thinking-machines, is transformed and multiplied, becoming a series of questions about the status of thought itself. Thus, Poe's question becomes not only, "What shall we think of thinking-machines?" but, more profoundly, "What shall we think of thought?" Moreover, where the human has already been identified with and as the function of thought, the question also becomes, "What shall we think of the human?" And, where, for Babbage, the trope of the thinking-machine has become not only the analytic of thought, but of natural law as well, the question finally becomes, "What shall we think of Nature/God?"

Thus, a number of conceptual shifts and reversals come in and out of play through Babbage's thinking-machines, so that thought slips between contingent and deterministic functioning, mechanism emerges as a signifier for *both* determinacy and reason, reason comes to signify both determinacy and contingency, and the concepts of the human and of mechanism intermittently overlap and converge. But, this play between deterministic "general" laws and the "unforeseeable" outcomes which they "develop"[48] describes not only the discursive effects of Babbage's engines, but also the material functioning of the mechanisms themselves. Thus, it is also the Bernoulli shift, or the baker's transformation, that Ada Lovelace, with Babbage's enthusiastic endorsement, will use in order to illustrate the "unlimited" capabilities of the Analytical Engine as a true thinking-machine.

Poe might rather have asked, "What does Mr. Babbage think of the calculating machine of Mr. Babbage?" If he had done so in 1836, he would have found several clues in the British press, including not only Babbage's own accounts of his work,[49] but also the detailed description of the first Difference Engine by Dionysius Lardner, written for the *Edinburgh Review* in 1834, which draws heavily from Babbage's own writing on the subject,[50] and which Babbage later cites approvingly in his own autobiography.

Extant evidence for American reports and reprints on Babbage's engines is copious, and, of course, the importation of British sources would have also contributed to Babbage's reputation within the United States. Certainly, by the time Poe writes his essay in 1836, he feels confident enough of his American audience's familiarity with Babbage and his engines to reference them casually, and to launch the part of his argument that relies upon the mathematical principles of the mechanism without further introduction or explanation. Several short notices of Babbage's initial success had appeared in American newspapers in 1822. But, beyond 1832 or so, Babbage tends to show up in American newspapers as "Mr. Babbage" or simply "Babbage," implying the writer's presumption of the audience's familiarity with the man and his work.[51] In 1841, Babbage is cited in the *New Hampshire Sentinel*, in a reprinted excerpt, as "Mr. Babbage, whose name is so well known among us as the author of the self-calculating machine" ("Extracts"). And by the time of Babbage's death in 1871, American newspapers and periodicals are not only providing detailed mechanical descriptions and histories of Babbage's calculating machines, but casually referencing both his name and his machines to variously signify genius, scientific marvel, and, significantly, unerring and unfeeling reason.[52]

Thus, one database of American journals turned up only four hits for Charles Babbage for the period from 1820-1840, but one hundred and eleven for the period from 1840 to 1890, during which time, Babbage and his calculating engines had become widely publicized in America. Thus, Poe's question, and the uncertainty about human identity in the context of the thinking-machine which it expresses, is reiterated in the American press, either directly or implicitly, throughout the remainder of the century.

Interestingly, Babbage's answers to these ontological questions are themselves largely responsible for the transformations they develop, and much of what appears in the American press can be traced ultimately back to Babbage's own texts.[53] The most extensive and comprehensive of these, Babbage's autobiography, published during the last decade of his life, is perhaps the most interesting for the way in which its narrative frequently treats the life of its author as the development of some natural law, or algorithm, responsible for its recurring pattern. Earlier and shorter accounts of the calculating machines and the principles behind them occur in scientific journals such as *Brewster's*, *Philosophical Transactions of the Royal Society*, and *Memoirs of the Astronomical Society*, as well as within his own volume on *Economy of Manufactures an Machinery*.

But, Babbage's *Ninth Bridgewater Treatise*, published almost thirty years earlier, was also widely covered in the American press, and, along with its controversial theory of miracles, also provides a general description of the functioning capabilities of the Analytical Engine. Philosophically oriented, the treatise speculates that the same deterministic rules that govern the contingent development of mathematical laws in Babbage's calculating engines might also be those general laws that govern the unfolding of the universe. Though evident in his more technical work, both the *Bridgewater* and the *Passages* texts self-consciously grapple with the apparent paradoxes in the play between determinacy and contingency, describing the calculating engines in terms of human thought processes such as memory and foresight--and so go a considerable distance toward answering Poe's question in ways that transform the conceptual terms of his question.

Babbage and Descartes
Charles Babbage writes in the early pages of his autobiography,
> [f]rom my earliest years I had a great desire to enquire into the causes of all those little things and events which astonish the childish mind. At a later period I commenced the still more important enquiry into those laws of thought and those aids which assist the human mind in passing from received knowledge to that other knowledge then unknown to our race (Babbage *Passages* 5).

Later, in the closing chapter of the same text, he reiterates this narrative framework for his life's work:
> I have always carefully watched the exercise of my own faculties.[...]
>
> Probably a still more important element was the intimate conviction that the highest object a reasonable being could pursue was to endeavor to discover those laws of mind by which man's intellect passes from the known to the discovery of the unknown (364-365).

These lines, while perhaps revealing more about the elder Babbage than about the younger,[54] recursively iterate the central concerns articulated by Descartes in both the *Discourse on Method* and the *Meditations*, as well as in numerous of his personal letters.

Like Descartes, Babbage makes conceptual and discursive

links between what he considers to be the proper methods of thinking and the natural functioning of the mind that verge on tautologies, bridging the concept of the human with that of reason, and reason with knowledge, but also knowledge with natural law, and, again recursively, that same natural law with mind.[55] Thus, in *Passages*, Babbage declares, "The great object of all my enquiries has ever been to endeavour to ascertain those laws of thought by which man makes discoveries" (340).

Like Descartes, the young Babbage apparently concluded that one such law involved skepticism toward received knowledge. In order to illustrate this point, he recalls an emblematic episode from early childhood precipitated by the ease with which he was able to trick a schoolmate into mistaking some shadows on the wall for ghosts. "[I]t naturally occurred to me, after some time," Babbage writes, "that as I had deluded him with ghosts, I might myself have been deluded by older persons" (Babbage *Passages* 7-8). According to his narrative, the young Babbage then embarks upon an inconclusive experiment in which he attempts to raise the devil in order to assuage his doubts about the latter's existence.[56] Following this incident, the young Babbage moved on from "doubts of the existence of a devil to doubts of the book and the religion which asserted him to be a living being" (Babbage *Passages* 9). Significantly, it was Babbage's own sense of justice which led him to those doubts, and

> to believe that it was impossible that an almighty and all-merciful God could punish me, a poor little boy, with eternal torments because I had anxiously taken the only means I knew to verify the truth or falsehood of the religion I had been taught (Babbage *Passages* 9).

Thus, through this narrative, Babbage points to a personal principal of skepticism concerning received knowledge and institutions, the origins of which he locates in his deepest childhood, and with which he identifies his lifelong intellectual project.[57] Through these themes of experimentation and introspection, Babbage characterizes himself as trusting only to those things that he can find within his own reason or experience—in other words, that which he perceives "so clearly and so distinctly that [he has] no occasion to doubt it" (Descartes *Discourse* 120).[58]

That is, in characterizing his intellectual quest in this way, Babbage casts himself as an heir to Descartes, and takes up the latter's intellectual (and, by extension, political and ethical) project of rejecting "anything of which [he] had been persuaded only by example and custom," instead "rightly using one's reason and seeking the truth in the sciences" (Descartes *Discourse* 116, 111). But, more specifically, Babbage also points to a lifelong interest in the mechanism of mind, that is, in the "laws of mind," which might be discovered through focused self-reflection (and, later, by experimentation with automata).

In this particular interest, too, Babbage's intellectual fascinations resonate with those of Descartes.[59] And, again, Babbage's autobiography provides a narrative through which to locate this interest in his earliest childhood.[60] While attending a mechanical exhibition of clockwork automata with his mother in Hanover Square,[61] Babbage's enthusiasm for the subject caught the attention of the exhibitioner, who offered the boy a private tour of his workshop. There, Babbage was enthralled by two unfinished automata not exhibited to the public, "two uncovered female figures of silver, about twelve inches high" (12). One of these was "an admirable *danseuse*, with a bird on the forefinger of her right hand, which wagged its tail, flapped its wings, and opened its beak" (12). The lady herself, he muses later in the same volume, "attitudinized in the most graceful manner" (273). Foreshadowing his later experience with his own computing automata, Babbage concludes the earlier passage by writing, "These silver figures were the chef-d'oeuvres of the artist: they had cost him years of unwearied labour, and were not even then finished" (12).

As an adult, Babbage would acquire the Silver Lady (as she was known to his friends) at auction after the artist's death. Before cladding her in the elaborate finery of the day, Babbage reports, he disassembled her completely and reassembled her in order to understand the details of her mechanism.[62] Thereafter, Babbage's famous soirées often featured the Silver Lady as the evening's entertainment, along with the unfinished section of the first Difference Engine, which was also displayed prominently in his home (Babbage *Passages* 273-274; Morrison xxiv).

Despite this childhood fascination with automata, Babbage's earliest computing interests had to do with systems of notation rather than with clockwork mechanism.[63] Indeed, during his time at Cambridge, Babbage's main passion was mathematics, an interest which he also traces to his earliest childhood (Babbage *Passages* 13-15). But, already at Cambridge, Babbage conceived problems of computing and logic as problems of communication and language.[64] Indeed, it was his somewhat earlier encounter with the idea of a universal language that Babbage credits with his ability to comprehend on his own the calculus notations of both Newton and Leibniz (Babbage *Passages* 18-19).

With this background, Babbage became embroiled in a political controversy involving the calculus while an undergraduate at Trinity College in Cambridge. Because British mathematicians were trained exclusively in the Newtonian system of notation, while the Leibnizian method was dominant in Europe, the top British mathematicians were often unable to comprehend the considerable Continental advances in the field. Thus, acting on the perceived inadequacies of his instructors (and, therefore, of his own education), and worried that British mathematics and science were becoming hopelessly marginalized and backwards, Babbage and a small group of friends[65] initiated a campaign for the use of the Leibnizian method in Britain, the success of which arguably helped to lay the groundwork for Britain's technological and scientific dominance over the next century (Babbage *Passages* 19-21; Hyman *Pioneer* 23-25).

Playing upon Leibniz's use of d's in his calculus notation

where Newton used dots, as well as referencing a contemporary controversy over efforts to print and distribute Bibles annotated for a popular audience, Herschel and Babbage eventually published a small volume entitled, *The Principles of pure D-ism in opposition to the Dot-age of the University* (Babbage *Passages* 21). Babbage also worked with Peacock and Herschel on a sort of text book of problems and examples, introducing the Leibnizian notation to maths tutors in a form that they would find both useful and seductive (Babbage *Passages* 28-29).

Thus, however much Babbage's narrative may seem to stress his feelings of intellectual isolation and self-creation, it is also clear from this same account how embedded in the intellectual problems of his moment Babbage was. Of course, the controversy over the calculus notation dramatically illustrates this point, and it is clear that Babbage greatly benefitted from the society of like minds in his friendships with Herschel and other Analytical Society members.

The Difference Engines
Though ostensibly a man of science, and, in the language of his time, a natural philosopher, throughout his career, Babbage was most driven to solve specific problems of technological engineering.[66] The most famous (and ultimately the most productive) of these problems was the challenge of calculating error-free mathematical tables of the kind used in maritime navigation,[67] a task at which he ultimately succeeded, despite the frustrating failure of the project as originally conceived.

Babbage the autobiographer locates the catalysts for this project as a nexus of his early interests in mathematics, in the related laws of reason and of language, and in clockwork automata,[68] tracing his first inkling of the idea to 1812 or 1813, while he was still at Cambridge, keeping company with the Analytical Society, and steeped in the controversies over the calculus notation (30-31). This episode, related by a fellow member of the Analytical Society, and not remembered by Babbage himself, is nonetheless the one he chooses to relate in his autobiography. In an earlier account, however, Babbage locates the origin of his idea of a calculating engine around 1820 or 1821. Meeting one night as a committee appointed by the Astronomical Society to prepare some tables, Babbage and Herschel sat down to compare the calculations prepared for them by two different computers.[69] "Finding many discordancies," Babbage writes, "I expressed to my friend the wish, that we could calculate by steam, to which he assented as to a thing within the bounds of possibility" (qtd. in Hyman *Pioneer* 49).[70]

There are two important points to be made about this latter account. The first of these is that Babbage stresses the *conceivability* of the project. He reports Herschel's response neither as dismissive, nor as incredulous; Herschel responds apparently matter-of-factly, "as to a thing within the bounds of possibility." This is in sharp contrast to Descartes' strong denial of the very *conceivability* of such a machine. Doron Swade suggests, however, that the proposition was not *widely* conceivable to Babbage's contemporaries and colleagues. In fact, Swade concludes, "expert opinion in England and on the Continent resoundingly rejected the utility of the engines for the purpose of tabulation" ("The Shocking Truth"). But, David Brewster's Babbage-influenced account of Babbage's Difference Engine also raises the issue of conceivability: "many of the most intelligent of our readers will scarcely admit it to be possible," he writes, "that astronomical and navigational tables can be accurately computed by machinery," and insists, based on his own observations, that "[a]ll this, however, Mr. Babbage's machine can do" (291).

In a similar passage, Dionysius Lardner makes the same point:

> To bring the practicability of such a project within the compass of public belief was not easy. [. . .] It transcended the imagination of the public in general to conceive its possibility; and the sentiments of wonder with which it was received, were only prevented from merging into those

of incredulity, by the faith reposed in the high attainments of its projector (Lardner 164).

As we have seen, the conceivability of a thinking-machine destabilizes the mutual relationships within a constellation of component and oppositional concepts including the machine, the human, and thought itself. Because these accounts of the Difference Engine were reported and reprinted in several U. S. newspapers and periodicals, they raise the issue of conceivability--and the anxieties about human identity that it opens--for an American audience.[71]

It is also important that Babbage's motivation in both accounts of his original idea for the Difference Engine is expressly characterized as the desire to introduce determinacy into what had hitherto been a highly contingent process. In this case, Babbage discursively and conceptually links contingency to the human, and it is the quest to remove "human error" from the calculating and recording process that provides the impetus for Babbage's idea. This idea is repeated in a handful of letters to Davy, Brewster, and the Astronomical Society in 1822, and later in the *Economy of Manufactures,* referencing the labor-intensive efforts of the French government's recent longitudinal tables project, and connecting the calculating engine's accuracy to its removal of human calculators and compositors.

A secondary problem with contingencies in the production of mathematical tables involved the process of their *re*production. Not only did errors occur in the initial calculations, but each time the tables were printed, new errors in copying could occur during transcription, typesetting, and proofreading. In a random survey done by Dionysius Lardner at London University, in order to prove the point, 3,000 known errors were found in the published errata sheets for forty volumes of a private collection. This did not include as-yet undetected errors; nor did it include the undetected errata bound to exist within the errata sheets themselves (Lardner 175-180; Swade *The Difference Engine* 12-16).[72] This association between the human and error, or contingency, is also repeated by Brewster, and thus disseminated through the American press. Quoted in *The New York Weekly Messenger,* Brewster writes that the engine's calculations will remain "absolutely free from error" because they "can be printed off without the aid of human hands, or the operation of human intelligence" ("Mechanics Department").[73] Thus, while there had been myriad attempts to remove error from the production of mathematical tables over the course of centuries (Swade *The Difference Engine* 12-16), it is Babbage who specifically imagines eliminating *human error*.

But, "human error," of course, is the result of contingency, and specifically of that contingency, which, we have seen, defines the human *as* thought for Descartes. Indeed, the machine's inability to perform contingently is precisely why a thinking-machine is inconceivable for Descartes. Thus, within this conceptual schema, there is something redundant about the phrase "human error," and it is precisely the function of thinking that introduces both error and contingency into the process of calculation. Babbage's proposal for the first Difference Engine, then, amounts to a plan for *the removal of contingency from the function of thought*, and poses a variety of conceptual problems for any simplified Cartesian understanding of human essence or identity.

Nevertheless, with Babbage's first Difference Engine, the trope of the thinking-machine begins to discursively associate reason and logic with the *non*-human, and to shift the trope of reason itself, so that it tends to signify determinism rather than contingency, as it did for Descartes.[74] However, in yet another reversal, Babbage will spend decades working on engineering solutions for bringing contingency into his machines' thinking processes. Babbage's own musings on the implications of the play of determinism and contingency in his calculating machines, as articulated in *The Ninth Bridgewater Treatise* and *Passages from the Life of a Philosopher*, are subtle, and anticipate the cybernetics and complexity theories of the twentieth century.

The idea of a calculating machine was, of course, not new with Babbage. Most notable among earlier devices were the calculators of both Leibniz and Pascal.[75] Both Leibniz and Pascal had built mechanical calculators with systems for carrying tens during the seventeenth century, and numerous calculating machines had been subsequently built and sold commercially. However, all required frequent intervention from an operator who understood the mathematical function being calculated, and thus neither of these mechanisms represented a calculating engine of the type that Babbage envisioned (Swade *The Difference Engine* 10-11; Hyman *Pioneer* 47-48).

Babbage's imperative was to eliminate human intervention once the initial "adjustments" had been made:

> Calculating machines comprise various pieces of mechanism for assisting the human mind in executing the operations of arithmetic. Some few of these perform the whole operation without any mental attention when once the given numbers have been put into the machine.
>
> Others require a moderate portion of mental attention: these latter are generally of much simpler construction than the former, and it may also be added, are less useful (Babbage *Passages* 30).[76]

Babbage goes on to say that, of the many machines that he has examined, none of those falling in the former category have been capable of operations beyond simple addition and subtraction. But, Babbage was proposing a machine that would not only do simple arithmetic, but also solve the complex polynomial equations required for nautical and actuarial tables, as well—all without the intervention of a human mind.

In other words, Babbage recursively iterates Descartes' concept of the human, linking it to its component concept contingency, in opposition to the machine and its component concept determinacy. But, these tropes, the machine and the human, while apparently signifying determinacy and contingency respectively, also begin to destabilize their opposition insofar as the function or activity to which they refer is understood as an operation of thought. With each iteration, or fold of the dough, contingency and determinism shift positions with respect to the human and the mechanistic, destabilizing their necessary, or essential, connection to either.

Babbage's solution to the problem of human error involved in calculation was a mathematical one based on the laws of logic and thought that so fascinated him. Rather than designing a machine to calculate each number in a table separately, Babbage took advantage of the regularities in certain polynomial equations, and designed a machine that could be programmed to follow the "rules" of any such table. Thus, using this already well-known "method of differences," in conjunction with several technical innovations for carrying tens and remembering previous results, Babbage was able to construct mechanisms that calculated complex operations using only the principle of addition (Babbage *Passages* 30-50; Hyman *Pioneer* 48-50; Swade *The Difference Engine* 28-31).

Babbage explains the system in *Passages* using a series of examples. In the first example, a table for the price of meat may be calculated by either multiplying the number of pounds by the price per pound for each number of pounds, or by starting with one pound and the price per pound, then adding one price per pound for every pound increase in the pound column. When using the second method, the entire table can be checked by calculating the last row of the column, using the first method. If the results agree, the entire table is known to be correct.

In an example using a progression of groups of marbles arranged as triangles, the rules governing the sequence of marbles is more complex, and requires a series of calculations based on previous results, or differences.[77] In other words, subsequent operations are *contingent* upon the results of previous operations. Nonetheless, each operation can be reduced to a series of additions or subtractions (the latter being a species of addition both mathematically and mechanically) based on fixed rules (35-42). "Any machine, therefore, which could add one number to another, and at the same time retain the original number [. . .] would be able to compute all such tables" (37). In some sense, then, the Difference Engine was not only a model, or an iteration, of the function of thought, but an analytic of the laws that govern it.

Babbage's solution to the second human error problem, that of reproduction, was to integrate the printing (or, in some versions of the design, the plate-making) process into the calculating machinery, so that an independent check of the final calculation in the series would also serve as the final printing proof. Babbage experimented with several systems, including a system for mechanically inserting moveable type onto a tray, the imprinting of a plaster of Paris mould using a type wheel, and a mechanism for punching copperplate with a steel punch. The latter became a machine tool in Babbage's own shop, and the device upon which many of the parts for the Difference Engine were machined (Babbage *Passages* 32-34). In each case, the primary goal was to make the printing process a determined outcome of the calculation process, and to eliminate entirely human error through the elimination of human intervention.

Thus, the first Difference Engine, as originally conceived by Babbage, is significant for its ambition of mimicking and *improving* not only the physical (mechanical) functions of the human body, but the human function of thought, or reason, as well. This language is reproduced not only in Brewster and Lardner, but in multiple reports appearing in U.S. newspapers and magazines.

Again, this discursive association between the human and error, or contingency, is significant, because, within a Cartesian schema, it makes the effect of the Difference Engine tantamount to removing the human (that is, contingency expressed as thought)[78] from the human, and begins to explain the anxious feelings of the uncanny that attend literary images of the thinking-machine throughout the nineteenth century. Thus, while machinery that could function faster or more precisely than the human body was not new, a machine that could think more efficiently and more accurately than the human mind was not only new, but also raised significant questions about the nature of human identity, and about its relation to its identifying component concept thought. It is really this new problematic of sorting out and attempting to realign the component concepts of the human and the machine that will characterize subsequent iterations of the thinking-machine in nineteenth-century America, and ultimately inform the fields of cybernetics and AI.

On July 3, 1822, Babbage sent a letter to Sir Humphry Davy, then president of the Royal Society, describing his plans for a calculating mechanism, announcing the completion of a small working model, and requesting financial backing for completion of a full-scale Difference Engine, which was partly granted. Babbage's original estimate of the time needed to complete the first Difference Engine was only two or three years, but the machine was still uncompleted in 1842, when the government withdrew all support for the project. Babbage, who had effectively stopped work in 1833, released the completed portions of the machine to the British government. This action was intended in part to counter a popular perception that Babbage had been receiving regular payments from the Crown while essentially drawing out a twenty-year boondoggle.[79]

In fact, the delays had mostly to do with disagreements and disputes with his contracted machinists--most notably with Joseph Clement, who stopped work at one point for over a year, then in 1833 stopped work entirely[80]—and the government's own wavering reluctance to fully fund the project. Babbage, in fact, lost a considerable amount of both time and money on the project, and felt deeply betrayed and ill-used by the Royal Society and the Crown. It must be noted, however, that Babbage's constant reworking of the plans also contributed to the delays, and by the time he gave up the project, he was already well into designs for a second, greatly improved Difference Engine, and the still more impressive Analytical Engine (Nicolas qtd. in Babbage *Passages* 52-74; Swade *The Difference Engine*

32-49, 55-71, 91-154).[81]

However, neither the second Difference Engine nor the Analytical Engine received either the financial backing or the public attention enjoyed by the first Difference Engine. Reluctant to risk more of his personal fortune, and even more reluctant to seek government backing a second time, Babbage made a few small models, but neither machine was completed during his lifetime (Science Museum 2; Hyman *Pioneer* 231).[82]

The second Difference Engine, like the first, operated on the principle of finite differences, but capitalized on "all the improvements and simplifications which years of unwearied study had produced for the Analytical Engine" (Babbage *Passages* 75). Uncharacteristically, Babbage left a complete set of drawings for this second Difference Engine, and completed the full design within a period of three years (Swade *The Difference Engine* 174). Upon the suggestion of the Earl of Rosse, Babbage did offer the drawings to the British government on condition simply that they build a working machine, but this proposition was rejected by the Prime Minister, Lord Derby, as being "indefinitely expensive" (qtd. in Swade *The Difference Engine* 176; qtd. in Babbage *Passages* 81).

The Analytical Engines, and the Difference of Contingency
Contemporary scholars routinely divide their discussions of Babbage's calculating machines between the Difference Engines and the Analytical Engines, citing the latter's capacity to function contingently not merely as a significant improvement upon the earlier devices, but as a category leap in the history of computing. The general consensus is that, whereas the Difference Engines were designed to compute a very specific type of mathematical equation according to predetermined rules (finite differences), the Analytical Engines were designed to process any problem involving any "objects [. . .] whose mutual fundamental relations could be expressed by those of the abstract science of operations" (Lovelace 270). Babbage himself writes of the Analytical Engine that "any directions may

be given which the circumstances require. [. . .] Any number of courses may be possible at the same time; and the choice of each may depend upon any number of conditions" (Babbage *Passages* 101). Choice, of course, is the very definition of contingency.[83]

Despite real advances in the Analytical Engine's capacity to respond contingently, and its generalized if-then logical approach to problem solving, it is nevertheless an oversimplification to draw too sharp a line between the Analytical and the Difference Engines. Indeed, the first Difference Engine incorporated several layers of contingency into its routine operations,[84] and prompted Babbage's initial philosophical musings on the relationship between natural law and miracles.[85] Furthermore, it was Babbage's work on the first Difference Engine which led directly to his work on the Analytical Engines, and recursively, his work on the Analytical Engine designs which prompted him to redesign elements of the Difference Engines in order to better deal with contingency (Hyman *Pioneer* 164-172; Swade *The Difference Engine* 95-125, 172-176; Babbage 45-47; 75-86).

The central nexus around which thought, contingency, and mechanism converged for Babbage also presented his greatest engineering problem from his first work on the Difference Engine, and spurred the insights which led to the Analytical Engine's generalizable functionality.

Babbage describes the problem in *Passages*: "unfortunately, there are multitudes of cases in which the carriages that become due are only known in successive periods of time" (45). Babbage refers here to the process of carrying tens in problems of addition. Since each carriage operation will take some specific unit of time, the machine will gain orders of speed if it can "remember" and "store" carriages to be performed in one operation. "Multitudes of contrivances were designed, and almost endless drawings made, for the purpose of economizing the time and simplifying the mechanism of carriage" (46). Both accuracy and speed obsessed Babbage in his quest to build machines that could outperform the human mind in mathematical thinking. In short, the problem had to do with how to fold the results of earlier calculations into subsequent ones; that is, how to store and retrieve the machine's contingent self-generating inputs into its ongoing processes. For Babbage, the solution to this engineering problem could be found in the laws of mind, and the challenge described as that of building "the operation of the faculty of memory" into mechanism (46).

Memory, of course, is a mode of thought,[86] and Babbage routinely describes his engines as performing functions of thought, that is, as thinking-machines. This is because, for Babbage, the rules of calculation *are* the laws of thought, and this conception of the calculating engines always points to their capacity to function contingently.

Doron Swade makes these connections clear in his comments regarding the first Difference Engine:

> Babbage's machine is the first known calculating device to successfully embody mathematical rule in mechanism. [. . .] you could for the first time achieve results that up to that point in history could only have been arrived at by *mental* effort—thinking. It was the first successful attempt to externalize a faculty of thought in an inanimate machine. [. . .] it is the first surviving artifact in the field of machine intelligence (*The Difference Engine* 83).

Again, though the Analytical Engine carries the principle of contingency in mechanism much further,[87] the first Difference Engine embodies that same principle, and serves as the initial basis for Babbage's reflections on "natural religion" and the theory of miracles, most fully elaborated in *The Ninth Bridgewater Treatise*.

Thus, even the "complete portion of the incomplete" first Difference Engine as described by Babbage embodies in mechanism the principle of contingency, and "can be employed to illustrate those singular laws which might continue to be produced through the ages, and then to be superseded by new laws" (Babbage *Passages* 49). And, again, it is Babbage's musings on the mechanical solutions to making the Difference Engine function contingently that lead him on the one hand to his philosophical work on the relationship between determinism and contingency, and on the other, toward the more universal Analytical Engine:

> The circular arrangement of the axes of the Difference Engine round large central wheels led to the most extended prospects.[88] The whole of arithmetic now appeared within the grasp of mechanism. A vague glimpse even of an Analytical Engine at length opened out, and I pursued with enthusiasm the shadowy vision (Babbage *Passages* 85).

Having "exhausted" the principle of successive carriage, that is of building the function of memory into his calculating engines, Babbage concludes that "it might be possible to teach mechanism to accomplish another mental process, namely--to foresee. [. . .] the next step was to teach the mechanism which could foresee to act upon that foresight" (Babbage *Passages* 46). This would be one of the principle accomplishments of the Analytical Engine.

Significantly, Babbage's language here ignores entirely any presumed essential opposition between the human and mechanism, but rather stresses the "human" functions associated with thought. Indeed, it seems to be an analytic of those processes and the laws that govern them. It is more useful, therefore, to think of the dynamic between the Difference Engines and the Analytical Engines as the recursively iterated problem of contingency in relation to the deterministic "laws of thought" (Babbage *Passages* 340) that had fascinated Babbage since childhood, and which he saw at work in the mathematical operations of his machines.[89] Significantly, this language of thought is repeated in numerous magazine and newspaper accounts, occasionally with more drama than Babbage himself employed.[90]

With Descartes, Babbage imagined a natural world that ran like clockwork, according to the fixed and determined laws established by its creator at the moment of creation.[91]

But, there is something very different in Babbage's conception of these laws, which, though fixed and determined, seem *themselves* to incorporate principles of contingency. Thus, Babbage writes that

> the most extensive laws to which we have hitherto attained, converge to some few simple and general principles, by which the whole of the material universe is sustained, and from which its infinitely varied phenomena emerge as the necessary consequences (Babbage *Bridgewater* 32).

And, while Babbage writes of an "omnipotent" Creator, who foresees all contingencies and necessary outcomes of his own design (Babbage *Bridgewater* viii-ix; 25),[92] implying a closed system; his central examples, the calculating engines, pivot on the *inability* of their creator to foresee the outcomes of their initial operational laws, and indeed, counts this as the main indicator of their utility (Babbage *Passages* 30). Thus, Babbage describes his relation with the Analytical Engine as involving contingencies beyond its creator's ability to anticipate:

> I had determined to invest the invention with a degree of generality which should include a wide range of mathematical power; and I was well aware that the mechanical generalisations I had organised contained within them much more than I had leisure to study" (Babbage *Bridgewater* 97).

Extending this principle of "generalisation" to Natural Religion, Babbage writes, apparently paradoxically, of a creator who is "intimately cognizant of the remotest consequences of the present as well as of all other laws" and yet whose human creation, "should require no future intervention to meet events *unanticipated by its author*" (Babbage *Bridgewater* ix; emphasis mine).

This apparent contradiction cannot be resolved where determinacy and contingency are conceived in opposition to one another, but Babbage's struggle within and against this conceptual and linguistic order begins to articulate the concepts that will emerge within theories of cybernetics, AI, and complexity, which understand determinism and contingency not so much as opposing but as mutually dependent, co-creating forces.

Thus, Babbage writes (and therefore thinks) within a conceptual framework inherited from Descartes, Galileo, and Newton, which tends to lead him back to the closed system, but he seems nevertheless always to be grasping to articulate a concept of the open, self-organizing, or complex, system. The difficulty for Babbage appears to be a directional error with respect to time, which prevents him from fully analyzing his own insights drawn from the calculating engines. When he writes that "we are irresistibly led, when we contemplate the natural world, to attempt to trace each existing fact [. . .] to some precontrived arrangement, itself perhaps the consequence of a yet more general law" (Babbage *Bridgewater* 31), he places necessity in the past, tracing the rhyzomatic branches of becoming back to some originary virtual and the moment that initiates their actualization.

This might be consistent with contemporary complexity theory. However, when he faces forward, Babbage tends to bring necessity with him, so that the unfolding bifurcation of becoming is also somehow ultimately predictable, at least by an omniscient creator. Thus, again,

> the most extensive laws to which we have hitherto attained, converge to some few simple and general principles, by which the whole of the material universe is sustained, and from which its infinitely varied phenomena emerge as the necessary consequences (Babbage *Bridgewater* 32).

In reading Babbage, it is also important to distinguish between real contingency (that is, undecidability) and mere epistemological uncertainty, something Babbage's biographers do not always clearly do. Indeed, Babbage's first illustration in the *Bridgewater Treatise* appears to point to problems of epistemology rather than to real contingency. He asks his reader to imagine a machine capable of effecting "computations of great complexity," and an observer sitting down before it (34). The machine is adjusted by the operator and set in motion, presenting a succession of figures to the observer at regular intervals. These numbers so presented represent the series of natural numbers (1, 2, 3, 4, 5, and so on) through to one hundred million and one. Based on his extensive observations, the observer deduces that the general law governing the operation of the machine is the addition of one to each successive number. However, rather than presenting the expected one hundred million and two as its next result, the machine shows one hundred million ten thousand and two, and thereafter a series of numbers known as the series of triangular numbers. The observer witnesses a succession of similar apparent changes in the laws governing the machine, yet without observing any intervention by the operator.

    The explanation Babbage offers is that each change in the apparent governing law "was a *necessary consequence* of the original adjustment, and might have been as fully foreknown at the commencement, as was the regular succession of any one of the intermediate numbers to its immediate antecedent" (39). Babbage then imagines a similar machine on an enormous scale, where

> a series of laws, each simple in itself, successively spring into existence, at distances almost too great for human conception. The full expression of that wider law, which comprehends within it this unlimited sequence of minor consequences, may indeed be beyond the utmost reach of mathematical analysis (42).

Extrapolating his observations of this theoretical calculating engine and its relation to it programmer and governing laws, Babbage concludes, "we can scarcely avoid remarking the analogy which they bear to several of the phenomena of nature," including evolution and geological change (43). Yet, again, despite the inability of human beings to grasp the general law behind all this complexity, Babbage returns to necessity and "the inevitable consequences of some more comprehensive law impressed on matter at the dawn of its existence" (49) which were "foreknown by their Author" (44).

In other words, "[t]he Being who called into existence this creation [. . .] must have known and foreseen all, even the remotest consequences of *every one* of those laws" (60). And, "all the combinations and modifications of matter can be supposed to be traced up to one general and comprehensive law, from which every visible form, both in the organic and inorganic world flows" (50).

In this illustration, Babbage describes a classic clockwork universe unfolding predictably (at least in principle) from initial inputs which "comprehend" and "embody" the entire scope of its unfolding. Thus, Babbage suggests, in his defense of miracles,[93] even an event as improbable and as unprecedented as a human resurrection might be plausible if it could be understood as "the exact fulfillment of much more extensive laws than those we suppose to exist" (92). Here there is no real contingency in that unfolding, only epistemological indeterminancy from the point of view of an ignorant observer. (But, even here, there is a contradiction between this ultimate knowability and that which "may indeed be beyond the utmost reach of mathematical analysis.")

Elsewhere, however, Babbage's language suggests that he glimpses something more radical in the play between contingency and determinacy. This mutual relation of determinism and contingency within the processes of one another is suggested vaguely in his "Reflection on Free Will." Here Babbage again uses the calculating machine in order to get at this dynamic. Just as "it is possible so to adjust the engine, that it shall change the law it is calculating into another law, at any distant period which may be assigned," so, too, it is possible that "this change may be made to take place *at a time not foreseen by the person employing the engine*" (*Bridgewater* 168; emphasis mine). That is, the change may be contingent upon the kind of result obtained by preceding calculations. But, he speculates, this kind of result, such as a number ending in seven when the operating law is the production of squares, may only be possible through preprogrammed moments of decision, and that, indeed, "the probability of any law with which we have become acquainted by observation being part of a much more extensive[94] law, of its having, to use a mathematical term, singular points or discontinuous functions contained within it, is very large (*Bridgewater* 150-151). In a letter to David Brewster, Babbage speculates further, that "although I do not determine the analytical law, I can produce the numerical result which it is the object of that law to give" (Babbage "Theoretical Principles" 310). That is, *the determinate operating laws of thought, nature, and the universe, may themselves incorporate functions of contingency.*

Later, in his autobiography, Babbage seems to be much clearer about the relationship between determinate laws of thought and contingency in the operation of his machines, as well as in thinking generally. Interestingly, Babbage's typical originary moment narrative on this issue references a series of chess games starting in his first year at Cambridge. One particular opponent who "studied chess regularly several hours each day, and read almost every treatise on the subject [. . .] took lessons from every celebrated teacher, and played with all the eminent players on the continent" gave Babbage a particularly hard time with his game (*Passages* 26). Babbage found that if he opened with a move predictable enough to have been written about, he invariably lost the game. His only hope of winning, he found, "was by making early in the game a move so bad that it had not been mentioned in any treatise" (*Passages* 26). In other words, Babbage found that his friend, whose play tended to merely reproduce past games and rely on known outcomes, could not respond effectively to the level of contingency introduced by Babbage. Thus, like Descartes, Babbage began to think that the laws of thought, though themselves determinate, must necessarily produce contingency, or function contingently.

But, rather than understanding this as the boundary between the human and the machine (or animal),[95] Babbage took it as a lesson on how to make his machines think more generally and flexibly--that is, better. Thus, in an important (and apparently unconscious) reversal of his initial goal of removing human error from calculation, building "contingency" into his calculating machines becomes Babbage's principle object for both the Difference and the Analytical Engines.

Babbage solved the engineering aspect of this problem by adapting the punch card system developed by Joseph Marie Jacquard for weaving cloth.[96] The cards rendered each Jacquard loom "capable of weaving any design which the imagination of man may conceive" (Babbage *Passages* 88). But each set of cards could also be made to produce a variety of designs depending upon the threads used for each element in the pattern. Similarly, the Analytical Engine consisted of the "store," or memory, containing all the initial variables as well as contingent variables produced through preceding operations; and the "mill," which operated upon the variables as they were fed into it at the appropriate time. Thus, the Analytical Engine required "two sets of cards, the first to direct the nature of the operations to be performed [. . .] the other to direct the particular variables on which those cards are required to operate" (*Passages* 89). The machine, then, punched its own results onto new cards, which were either stacked as final results, or held in the store until needed for further operations. Thus, the "Analytical Engine [was] therefore a machine of the most general nature," and, "in principle, any operation reducible to this form of notation could be "developed"[97] (*Passages* 89).

Indeed, Babbage often juxtaposes claims for the contingent nature of the Analytical Engine and formulations of its activities using discursive formations normally associated with human thought processes. For instance in one explanation of the machine's mechanical principles, he enumerates "eight conditions, each of which is absolutely unlimited as to the number of combinations" (*Passages* 93). These include the number of constants, the number of variables, and the number of functions possible for any given program, extending the machine's analytical power to "infinity" (*Passages* 93).

Elsewhere, Babbage writes of "maturing an engine of almost intellectual power" (qtd. in *Passages* 80).[98] In the same letter to Lord Derby, Babbage introduces the Analytical Engine as having "over the most complicated analytical operations [. . .] nearly unlimited power" (qtd. in *Passages* 79). He variously describes the Analytical Engine as needing "intellectual food" (*Passages* 92), economizing "intellectual labour" (qtd. in *Passages* 81), and performing "mental tasks" (*Passages* 134). As to the moment-to-moment operation of the engine, Babbage writes, "Any number of courses may be possible at the same time; and the choice of each may depend upon any number of conditions" (*Passages* 101). In these cases, the Engine is capable of "the act of judgment sometimes required during an analytical enquiry, when two or more different courses presented themselves [through] the usual,[99] but very tedious proceeding of approximation" (*Passages* 98-99).

This last point Babbage elaborates with a reference to the Duke of Wellington, and his skill as a decision-maker on the field. Like Wellington, the Difference Engine could be faced with "almost *innumerable* combinations [. . .] a number so vast that no human mind could examine them all" (*Passages* 133). Writing about the Difference Engine through a flattering account of Wellington, Babbage describes the process of thought involved in determining an outcome:

> [. . .] he must also be able to foresee. [. . .] he must undertake one of the most difficult of mental tasks, that of classifying and grouping the innumerable combinations. [. . .] he must be able to discover, to fix his attention, and to act upon the most favourable. Finally, when the course selected [. . .] is found to be entirely deranged by one of those many chances inseparable form such operations, then, in the midst of action, he must be able suddenly to organize a different system of operations, new to all other minds, yet

[...] anticipated by his own.
The genius that can meet and overcome such difficulties *must* be intellectual. [...] (*Passages* 134-135).

Thus, while the programmer may not be able to foresee the results of his or her programming, it is the *machine's* ability to "classify," "organize," "fix attention," and "foresee" that makes its function indistinguishable, at some level, from human thought. But, perhaps most dramatically, Babbage writes that the "Difference Engine is, like its prouder relative, the Analytical, a *being* of sensibility, of impulse, and of power" (*Passages* 83; emphasis mine).

This tension between the deterministic functioning of mathematical rules embodied in and governing mechanism, and the contingent functions of mechanized thought are not only at play in Babbage's own writing, but as well in the work of his closest contemporary commentators, General L. F. Menabrea and Ada (Byron) Lovelace.[100] For example, in a passage that suggests a roughly Cartesian discursive formation, Menabrea writes that "the engine, from its capability of performing by itself all these purely material operations, spares intellectual labour," freeing the human mind of genius for meditation (266). But, in order for material operations to spare intellectual labour, the intellectual labour must itself be material--and that is *not* the Cartesianism which Menabrea seems so reluctant to abandon.

Menabrea continues to struggle within an inadequate language and conceptual framework,[101] writing that the Jacquardian punch cards "are able to reproduce all the operations which the intellect performs in order to attain a determinate result," and that the machine cannot "interpret a result not of itself evident, since it is not a thinking being," but then immediately proceeds to describe what actions take place once the machine has "foreseen" a result (263).

Earlier, Menabrea writes that Babbage's Analytical Engine, "so as to embrace the solution of an infinity of other questions included within the domain of mathematical analysis," embodies "all the generality of algebraic notation" (251). That is, he describes an automaton at least theoretically able to "act in all the contingencies of life in the way in which our reason makes us act" (Descartes *Discourse* 140). But, also that "it must exclude all methods of trial and guesswork. [. . .] for the machine is not a thinking being, but simply an automaton which acts according to the laws imposed upon it" (Menabrea 252). Thus, Menabrea has described the Analytical Engine as both a contingently functioning mechanism and as an automaton ruled by deterministic laws within the space of less than two pages. But, in a typical gesture, Lovelace increases the complexity of this passage with another fold of the dough, adding in a note that

> This must not be understood in too unqualified a manner. The engine is capable, under certain circumstances, of feeling about to discover which of two or more possible contingencies has occurred, and of then shaping its future course accordingly (Menabrea 252).

Indeed, it is Lovelace who seems to have the better grasp of the Analytical Engine's implications and potential. Though she at one point writes that the Analytical Engine can neither "*originate* anything" nor do other than "what we *know how to order it* to perform" (Menabrea 300), she elsewhere suggests that "the engine might compose elaborate and scientific pieces of music of any degree of complexity or extent" (Menabrea 271).

Again, highlighting the scope of the Analytical Engine's potential, Lovelace describes the it as "the *material and mechanical representative* of analysis" (Menabrea 272) and as "the *embodying of the science of operations*" (Menabrea 270);[102] and then defines the unlimited "operations" which the Analytical Engine might perform as "*any process which alters the mutual relation of two or more things*" (Menabrea 269).[103]

Finally, Lovelace explicitly invokes Descartes, describing the Analytical Engine as "a uniting link [. . .] between the operations of matter and the abstract mental processes," so that "not only the mental and material, but also the theoretical and the practical [. . .] are brought into more effective and intimate connexion" (Menabrea 273). And, thus, she concludes, quite unequivocally, that the Analytical Engine represents the first "practical possibility [. . .] of a thinking or of a reasoning machine" (Menabrea 269).

In any case, whether Babbage ultimately saw the laws of thought as deterministic or as contingent (and the point to be made is that both Babbage's technology and his thinking on the subject problematized the opposition), it is clear that he and the people he most closely influenced, understood the calculating engines to embody and execute those laws, and in a very real sense, to think.

Communication as Control
It is also worth noting briefly that for Babbage the problems of mind, mathematics, mechanism, and signs were always closely intertwined. Indeed, it is fair to say that he understood both mathematical and engineering problems to be essentially problems of communication and language, and this association is expressed throughout his writing. Thus, he tells us in his autobiography that "the philosophy of signs always occupied [his] attention (*Bridgewater* 322), and that his "early perception of the immense power of signs in aiding the reasoning faculty contributed much to whatever success [he] may have had" (*Passages* 364).

Babbage tells us also that his interest in language and systems of signs began in early childhood, when he became expert at breaking the codes devised by other boys. As an adult, Babbage devised telegraphic codes based on mathematical principles for the British Government, and developed a system of maritime signals using flashing lights for the navy (Hyman *Pioneer* 226-228), as well as similar systems using "occulting lights" for communicating to a ship its distance from a particular lighthouse, and for identifying that lighthouse (*Passages* 245; 340-342). In his campaign at Cambridge over the dots and dashes of calculus, it is difficult to know whether it was the mathematics or the system of signs that most captivated him. Indeed, for Babbage, the two could not properly be separated.

Furthermore, in Babbage's own estimation, his mechanical notation, developed in order to aid his work on the Difference Engine, stands as "one of [his] most important additions [. . .] to human knowledge" (*Passages* 340). In his 1852 letter to Lord Derby, Babbage described the interdependence of his work on the calculating engines with this system of signs:

> I have invented and brought to maturity a system of signs for the explanation of machinery, which I have called the Mechanical Notation, by means of which the drawings, the times of action, and the trains for the transmission of force, are expressed in a language at once simple and concise. Without the aid of this language I could not have invented the Analytical Engine; nor do I believe that any machinery of equal complexity can ever be contrived without the assistance of that or of some other equivalent language. The Difference Engine No. 2 [. . .] is entirely described by its aid (qtd. in *Passages* 79).

This system, Babbage notes, is a *general* system capable of describing military battles or of "representing the functions of animal life" (*Passages* 109). But, at the same time, it is also a system for eliminating ambiguity, that is, contingency (or white noise). Thus, like the Analytical and Difference Engines, this encoding of thought, or analytics, also relies upon the play between determinism and contingency, which move around one another, and around the trope of the thinking-machine, in increasingly unpredictable ways. In this sense, the mechanical notation, a kind of programming, is also a thinking-machine.

But, Babbage was not just interested in the intrigue or utility of languages and systems of signs; rather he saw these as the basic underlying structures of all scientific and engineering problems. For example, during his work on the Difference Engine, Babbage devised a method for feeding the contingently calculated results back into the machine as input, thus eliminating human intervention, and therefore human error--that is, both removing and adding contingency to the system in one stroke. He described his solution as "a snake eating its tale" (Swade *The Difference Engine* 95)--a perfect description of the concept of feedback, which would become central to the field of cybernetics.

Babbage also writes of the Analytical Engine that "the law of its development must be communicated to it by two sets of cards" (*Passages* 89). This language is typical, and suggests that Babbage understood the punch card system appropriated for the Analytical Engine from the Jacquard loom, that is, the source of *both* the machine's determinacy and its *generality or contingency*, in terms of language and communication. Thus, Babbage saw his engineering solutions to mathematical and technical problems presented by the calculating engines themselves as forms of communication, the central problematic of twentieth-century cybernetics and AI.

Babbage's Chess-Player
Perhaps Babbage's most interesting work concerning thinking-machines had to do with the possibility of building a chess-playing automaton. Touted by himself as one of the "author's further contributions to human knowledge," he set forth the principles by which a machine might be constructed which "should be able to play a game of purely intellectual skill successfully" (*Passages* 349).

In a typically recursive gesture, Babbage characterizes the project as an attempt to observe his own intellectual process at work solving a new problem (*Passages* 349). However, Babbage not only had seen von Kempelen's chess-player during Maelzel's London tour in 1819, but had challenged it to a game (and lost) a year later (Standage 140). Babbage also tells us that he made "enquires into the relative productiveness of the various exhibitions of recent years," with an eye toward raising money in order to complete the Analytical Engine, but concluded that "the most profitable exhibition that had occurred for many years was that of the little dwarf, General Tom Thumb" (*Passages* 353), a clear (and ironic) reference to an erroneous exposé of the von Kempelen hoax. While Babbage did not believe that von Kempelen's chess-player was, in the words of Poe, a "pure machine," he did, through his work on the problem, come to believe that "every game of skill is susceptible of being played by an automaton" (*Passages* 350).

Babbage begins by polling friends and acquaintances as to whether they *believe* that games of skill require human reason. Most do, Babbage tells us, and even those best acquainted with mathematical science deny "the possibility of contriving such machinery on account of the myriad of combinations which even the simplest games included" (*Passages* 350). It is this same assumption, that an automaton can only move in a "fixed and determinate [. . .] progression" from initial data, and must therefore have each contingency built into it as a separate mechanism in order to process contingency, that had mislead both Descartes and Poe (Poe "Maelzel's Chess-Player" 349).[104] That is, for Poe and Descartes, contingency, as a sign of reason and mind, is still, therefore, a sign of the human; for Babbage, however, contingency, mind, and reason are no longer necessarily human, but merely material "developments" or processes (*Passages* 89). It is precisely for this reason that Babbage can conceive a thinking-machine.

Babbage solves the chess problem in an entirely new way, according, he believes, to the laws of thought. The machine does not need to know in advance all possible chess combinations, Babbage reasons. It need only know the rules of the game, and then all legal moves for the specific arrangement of the pieces for any given turn. Hence "the question is reduced to that of making the best move" for each possible combination of pieces (*Passages* 350). Thus, using the mechanisms for "memory" and "foresight" already designed for the Analytical Engine, Babbage's theoretical chess-player asks itself, and then acts upon, a series of questions arranged according to an if/then logic: is the arrangement of pieces consistent with the rules of the game?

if so, has the machine already lost? if not, has it won? if not, can move x win the game on the next move? if so, make the move. if not, can this move win in two subsequent moves? in three? and so on.[105] Using this logical map, and knowing the existing capacity of the Analytical Engine, Babbage concludes that, "the combinations involved in the Analytical Engine enormously surpassed any required, even by the game of chess" (*Passages* 350-351). Thus, Babbage's "proof" can be understood as yet another analytic of the human thought process, and as a coding of that analytic.[106] And, for Babbage, this means the fluctuating play between deterministic rules and their contingent expression.

After demonstrating how he believes a machine operating through a determined set of simple rules would be able to deal with an infinity of contingencies, and in order to streamline the problem, he sets chess aside, and considers the far simpler game of "tit-tat-to," using the same basic logic he had developed for a chess-playing Analytical Engine, but on a much smaller scale.

Babbage observes that, as with the Analytical Engine, "cases arose in which it became necessary [. . .] that the machine itself should select one out of two or more distinct modes of calculation" (*Passages* 352). But, the game-playing machine might be presented with an undecidable choice. That is, two or more choices might be equally conducive to winning the game. Babbage's solution to this problem is significant in that it introduces an entirely contingent element into the automaton's play. Designed to keep track of the number of games played since its creation, the machine will divide that number by the number of equally desirable moves. The results of this division will decide the dilemma. In other words, the machine's choice would be determined by the number of games it has played in its lifetime--*an element of pure chance* within the context of the game being played. Apparently, Babbage had not forgotten the lessons learned from his Cambridge chess partner, Brande.

But, his work with the chess-playing automaton begins

to explain the anxiety that will surround the trope of the thinking-machine in the American popular imagination during the nineteenth century. For there is a suggestion, at least, that if a machine could be made to play a game of chess, it could necessarily be made to win every game. That is, since each move is in a sense the "first move" for the remainder of the game, and the problem reduces to the machine making the best move at any turn, then "if the automaton could make the first move rightly, he must be able to win the game, always supposing that, under the given position of the men, that conclusion were possible" (*Passages* 350). In other words, a machine capable of the task of chess-playing, a task which until that point could only have been performed by a human mind, would necessarily perform that task *better than a human being*. That is, the machine would not only function as a human, but it would function as a *better human*.

We have already seen this implication in a more circumscribed manner with regard to the Difference Engine, whose entire motivation was to calculate algebraic tables more accurately and more quickly than could human calculators.[107] Of course, industry had for decades been replacing the physical laborer with more efficient and accurate machines (such as the Jacquard loom). But, now the Analytical Engine represented a "true manufactury of figures. [. . .] capable of aiding human weakness" (Menabrea 266). Thus, if the human is defined by its unique capacity for contingency, reason, and thought, and if these are all, in fact, roughly equivalent component concepts of the concept of the human, what happens to the human when the machine appropriates these functions and identifying concepts to itself? What happens to the human when the machine is *more* human?

Finally, then, we may read Gibson and Sterling's fictional Ada Lovelace as giving Babbage's answer to Poe's question. Certainly, a blurring of boundaries between the human and the machine is the result of the machine's appropriation of these functions and component concepts that define the human, and to the extent that it does so, it eliminates the human as a distinct ontological category. This can be seen clearly in Babbage's language describing the machine "foreseeing" and "judging," as well as in his use of the personal pronoun when discussing the proposed chess-playing automaton (*Passages* 350). Thus, in a sense, there is something like the Heisenberg Uncertainty Principle at work, so that as the machine becomes more clearly identified as human, the human itself becomes less certain and more vague.[108] And, so, in this conceptual sense, as well as in a very real material sense, Babbage's thinking-machine threatens to supplant the human.[109]

CHAPTER THREE/ THIRD ITERATION: THE AUTOMATON CHESS-PLAYER AND OTHER FICTIONAL THINKING-MACHINES

> "A thinking machine? *Was the Chess-Player such a creation? . . .* A thinking machine! *Was such a thing really conceivable?*" (Joseph Friedrich Frieherr zu Racknitz qtd. in Wood 61).

As we have already seen, in a world governed by Cartesian concepts of the human, the machine, and thought--that is, where the human is identified as contingency, the machine is identified as determinism, and thought is identified as being uniquely human by virtue of its being contingent-- a thinking-machine is strictly (that is, *a priori*, by definition) inconceivable. Furthermore, and for this reason, the conception of the thinking-machine, as the child of human thought, would necessarily entail the reorganization and redistribution of the concepts involved in human identity, placing the concept of the human itself in crisis.

It is within this dynamic that Charles Babbage, commencing in 1822, repeatedly conceived, and expressed the concept of, the thinking-machine in multiple iterations using text, "mechanical notation," wood, iron, bronze, and steel. It is also within this dynamic that the *image* of a chess-playing automaton appeared, eliciting both excitement and anxiety throughout two continents, and across social and economic classes, inspiring countless letters, articles, essays, and pamphlets explaining, defending, or denouncing its appearance as a "pure machine."[110] Of course, every letter or essay denying the conceivability of the thinking-machine participated in its conception, and the more various and multiple were its iterations, the more they chaotically disorganized the concept of the human while beginning to coalesce around the emergent concepts of cybernetics and the posthuman.

Kempelen's Chess-Player
Wolfgang von Kempelen first exhibited his chess-playing Turk for Empress Maria Theresa of Austria-Hungary and a small group of courtiers in 1770, in response to a self-given challenge to construct a machine "the effect of which would be much more surprising, and the deception far more complete" than the "conjuring" that he had been invited to witness and explain to the court (Standage 18-19).[111] The Turk immediately became a sensation in Vienna, where it performed frequently for the court, and news of the automaton spread rapidly throughout Europe, as letters from witnesses circulated privately and began to appear in foreign journals and newspapers (Standage 32).

It is fair to say that subsequent responses to the chess-player were nothing more (or less) than variations of the themes, arguments, details, and solutions offered in these early accounts. For example, one of the earliest and most widely circulated eyewitness reports of the chess-player's private performances was a letter written in 1770 by Louis Dutens to the editor of *Le Mercure de France*. Dutens describes the Turk[112] and its performance in some detail, and reports a variety of theories explaining its operation, including magnetism and the concealment of a small person (or monkey) within--both of which he discounted on the basis of his own observations, and both of which would persist for decades as popular explanations (Brewster 273; Standage 34-35). Though Dutens himself declined to suggest a plausible solution to the problem, he apparently believed the chess-player to be somehow controlled by von Kempelen. Nonetheless, the image presented by Dutens is not only that of a clever trick or slight of hand, when he writes of "an automaton which can play chess with the most skillful players" (qtd. in Standage 32). It is rather the image of a thinking-machine.

Moreover, it is, apparently, an image which excited a good deal of anxiety.[113] Dutens himself reports that "I have met several people who played chess neither as quickly, nor as well as the Automaton, but would even so have been greatly affronted to have been compared to it" (qtd. in Standage 33). One particular letter to *Le Mercure de France* in response to Dutens begins to indicate why this might be so: the chess-player, the writer argued, *must* contain within it a small child, since a machine could never be designed to function spontaneously (Standage 34). In other words, it is inconceivable that the Turk's deterministic mechanical body could be responsible for contingent play; that is, in order for it to function contingently, the body *must* contain a human mind.[114]

Though neglected by Kempelen for many years, after its initial tour, and eventually dismantled, the Turk was rebuilt at the behest of Maria Theresa's successor, Joseph II, and toured Europe with von Kempelen from 1783 until 1785 (Standage 40-42).[115] This tour took von Kempelen and the chess-player to Paris and London, where the Turk was publicly exhibited, accepting challenges from all comers, including prominent chess masters, as well as mere aficionados such as Benjamin Franklin (Standage 48ff; Wood 84).

Chess had been popular in Europe throughout the eighteenth century, but was particularly fashionable within high society and among the intelligentsia during the 1770s and 1780s, and both first cities were centers for the game.[116] In Paris, the Turk lost well-publicized matches with two first-ranked players, Legall de Kermeur and François-André Danicon Philador, while winning nearly every other match it played, and consistently drawing large crowds (Standage 43; 45; 49-53).

And, in Paris, again, the chess-player's performances precipitated debate along now-recognizable lines of reasoning. Louis Petit de Bachaumont, court chronicler at Versailles, declared the Turk to be "far superior to other previously-known automata, such as the digesting duck, the flute-player, etc. [since it] performs not just physical motions, but elevated intellectual functions" (qtd. in Standage 44). However, the *Journal des Savants* complained that the chess-player was "described quite overtly in certain newspapers as though it were an automaton that really plays, and on its own" (Standage 53). And, Melchoir von Grimm, writing to Diderot, echoed the letter-writer to *Le Mercure de France*, arguing that "[t]he machine would not know how to execute so many different movements, which could not be determined in advance, unless it was under the continual control of an intelligent being" (qtd. in Standage 46).[117] A short time later, as the chess-player toured London, Henri Decremps would suggest that the box upon which the Turk sat concealed a dwarf.[118] Philador, however, widely recognized as *the* best chess-player in Europe, apparently believed the automaton to be genuine, claiming that "no human opponent had ever fatigued him to the same extent" (Standage 52).

Late in 1783, von Kempelen took the chess-player to London, where they were similarly received.[119] The English translation of Carl Gottlieb Windisch's *Inanimate Reason* appeared in London at about the same time, and describes the Turk as "indisputably the most astonishing automaton that has ever existed" (qtd. in Standage 62).[120] His description of the automaton and its functioning is particularly interesting as yet another example of the way in which iterations of the thinking-machine confuse and rearrange the component Cartesian concepts of the human, the machine, and thought, so that even a quick glance at Windisch produces the confusing (or confused)-- and, one would expect, inconceivable--image of an automaton that is both reasoning and inanimate.[121]

But, despite the *Monthly Review*'s rebuke of those "simple enough to affirm, both in conversation and in print, that the little wooden man played *really* and *by himself*" (qtd. in Standage 62), the greater force of the response to the chess-player in London, as elsewhere, came in the once again familiar form of attempts to expose it as deceit. Thus, asserting the inconceivability of the thinking-machine within the Cartesian schema that links the body with the "automatic," and the mind with contingency, Philip Thicknesse wrote in his 1784 pamphlet, *The Speaking Figure, and the Automaton Chess-Player, Exposed and Detected*:[122]

> That an Automaton may be made to move its hand, its head, and its eyes, in certain and regular motions, is past all doubt; but that an AUTOMATON can be made to move the Chessmen properly [. . .] in consequence of the preceding move of a stranger [. . .] is UTTERLY IMPOSSIBLE" (qtd. in Standage 63; Wood 67).

Thicknesse went on to write that the Automaton certainly "is a man within a man" and "bears a living soul" within its outward form (Standage 66), offering the perfect image of Cartesian dualism. Thicknesse's specific solution to the operation of the chess-player involved the concealment of adolescent children within the mechanism, and mirrors; but these details are less interesting than the intensity of his apparent outrage at the *very idea* of a thinking-machine, and the way in which his argument is discursively structured as a defense of the kind of popular Cartesianism later critiqued by Ryle.

Von Kempelen returned to Europe where he toured with the Turk in 1784 (Standage 80). Texts attempting to understand the chess-player's operation continued to proliferate, and, again, these generally had the effect of shuffling and redistributing the component concepts of the human, the machine, and thought in interesting and surprising ways. Most notable of these was a book by Joseph Friedrich, Freiherr zu Racknitz, a rather rigorous assessment of the major theories regarding the Turk published to that date. On his way toward coming quite close to the actual solution of the Turk's operations, Racknitz defends the argument that the Turk could not be an actual automaton, since this would require every possible response to every possible move to be pre-calculated in advance, suggesting an image of the thinking-machine--and of thought itself--as thoroughly deterministic (Standage 83).[123]

Maelzel's Chess-Player
At the end of the Turk's tour in 1785, the chess-player was again packed away into its crates, where it remained until 1808, at which time it was bought from von Kempelen's estate by Johann Nepomuk Maelzel, himself an inventor and builder of automata. Maelzel, after making some of his own modifications, then selling and re-purchasing (or possibly leasing) the automaton years later, exhibited the chess-player once again throughout Europe (Standage 102-117).[124] In 1818, Maelzel moved the chess-player to Britain, touring London, northern England, and Scotland, and returning to London in the fall of 1819 (Standage 124).[125]

Yet another host of attempts to discover the operations of the automaton were prompted by this second tour of England, including Robert Willis's 1821 pamphlet, *An Attempt to Analyse the Automaton Chess Player* (Standage 128). It is Willis, most probably through Brewster, that forms the substance of much of Poe's essay on the chess-player, both technically and philosophically. Not only does Poe take nearly all the details of the chess-player's appearance, construction, and performance from this source, but (most probably) most of his "*a priori*" argument, and the technical solution as well.[126]

In 1826, fleeing the previous owner's attempts to collect

a debt still owed on the Turk, as well as the wrath of the inventor of the metronome that Maelzel had recently patented, Maelzel arrived with the chess-player in New York, and began a tour of the United States which lasted until his death in 1838 (Standage 150; 190-191).[127] That the chess-player's reputation preceded it is clear, and Maelzel's arrival at New York Harbor, along with "the long known, but not yet discovered, chess player, which has so long puzzled and surpassed all Europe," was reported by the *New York Evening Post* ("The Automaton Chess Player").[128] The text presents the chess player's operation as a mystery which has "long puzzled [. . .] Europe"; but this puzzle is framed as "the attempt to discover by what secret *springs* [emphasis mine] its movements are directed," suggesting it as given that the machine moves *itself*, "guided by an uncontrolled free will" ("The Automaton Chess Player").

Later that spring, a report of the chess-player's first public appearance in the United States was picked up and reprinted at least as far west as Sandusky, Ohio. This report is both typical and notable. In addition to the standard description of both the audience and the automaton, including details about the chess-player's display by Maelzel and its movements during the performance, the article struggles within and against a Cartesian conceptual framework to address the question of the conceivability of mind in mechanism.

> It is an old saying that 'men are machines,' but Monsieur Maelzel's automaton almost persuades us to reverse the axiom, and say *machines* are *men* [. . .] Knowing that the machine itself is but a combination of wheels and springs, we are astonished to witness in its movements the result of accurate thought and profound calculation. How this intellectual principle is conjured with the machinery [. . .] is beyond our power to imagine ("The Automaton Chess Player" *The Sandusky Clarion*).[129]

The article can be understood as "post-Cartesian" in its reference to La Mettrie's *L'Homme Machine*, and points to the conceptual confusion that Bierce will exploit (with specific reference to La Mettrie) in "Moxon's Master": that is, if humans are machines and machines function as humans, humans and machines become indistinguishable as conceptual categories. Furthermore, the text compounds multiple reversals of the Cartesian identification of thought with contingency, so that thought becomes "accurate" (suggesting a deterministic precision) and calculation "profound" (suggesting an almost spiritual quality to the automatic), while compounding thought and calculation themselves. This is precisely the confusion and redistribution of the component concepts of the human and machine that is evident in much of the writing on the Turk, including Poe's, and from which the concepts of cybernetics and the posthuman will emerge.

Similar accounts of the chess-player's exhibition, attended by flurries of "solutions" and refutations of the automaton's authenticity in editors' and letters columns, appeared in newspapers and journals over the course of the next twelve years.[130]

Furthermore, accounts of Maelzel's chess-player did not disappear even when he eventually stopped touring.[131] A typical essay appearing in *Atlantic Monthly* in 1858 touts the mechanical wonders of the age, while invoking Maelzel to suggest that, "The human mechanic must be content, if he can approach as near to the creation of life as the painter and sculptor have done" ("What Are We Going to Make?"). Toward the later part of the century, Maelzel and the chess-player persisted as tropes, often casually illustrative of some loosely related point, as in the description of an ancient book seller who repeats philosophical and scientific jargon with no understanding, "like Maelzel's Chess-Player" ("Editor's Drawer"),[132] or an account of a cigar wrapper whose machinery, "like the automaton chess player of Maelzel . . . is concealed in a box" ("American Institute Fair"). Furthermore, occasional stories about the chess-player were still running well into the twentieth century ("Miracles" 1922), indicating that the trope of the thinking-machine as iterated through Maelzel's chess-player resonated multiply and persistently within popular discourse. From the beginning, then, the pressing problematics attending the chess-player fell within two general categories. The first of these encompassed a broad set of how-does-it-work questions addressing the mechanical details of von Kempelen's (and later of Maelzel's) specific machine, e.g., is there a dwarf concealed within the body of the Turk?; does von Kempelen/Maelzel control the automaton's actions through the use of magnets or wires?; does the device play according to predetermined rules without human intervention? But, there was also, from the beginning, a more general and profound set of questions about what machines might conceivably (or not) be capable of doing.[133]

So it is that when Edgar Allen Poe published "Maelzel's Chess-Player" in 1836, he was participating in a conversation with a rich history. More than that, he was largely reproducing it, or, rather, reshuffling and multiplying it. Indeed, despite his inflated claim to provide, at last, final and absolute answers to the important questions surrounding the chess-player, there is scarcely anything new in the details either of Poe's technical solution or of his philosophical argument.[134] But there is something remarkable in Poe's essay nonetheless, namely an apprehension, an anxiety, an urgency of denial about this looming image of the thinking-machine. For it is not enough for Poe merely to demonstrate that *this* machine is not a "pure machine," that *this* chess-player hides a human mind.[135] Indeed, the real urgency of the essay seems to be focused on proving "*a priori*" the absolute *inconceivability* of a "pure" chess-playing machine, and, in fact, of a thinking-machine of any kind (Poe "Maelzel's" 319).[136] This latter is, in fact, Poe's first argument against the authenticity of Maelzel's chess-player as a pure machine, and despite his extensively detailed technical solution to the mechanical problem of accomplishing the illusion, he believes that his "*a priori*" argument renders such solutions redundant, or of secondary importance.[137] Moreover, upon this point, Poe admits of no debate. That Maelzel's chess-playing machine might be a pure machine is not an open question for Poe. That such a device *in principle* must require the intervention of a human mind is "susceptible of a mathematical demonstration, *a priori*. The only question then is the *manner* in which human agency is brought to bear" (Poe "Maelzel's" 319). Poe's argument, therefore, is two-fold. On the one hand, he attempts to demonstrate the specific details of the means of the fraud's execution. But, more importantly, and more problematically, he also offers what he believes to be a general proof for the *inconceivability* of a chess-playing machine.

Interestingly, Poe demonstrates this "mathematical" proof using Charles Babbage's Difference Engine.[138] But in terms of this argument, Poe has nearly as much to say here about the Difference Engine as he does about the chess-player, and it seems to be as important to him to discredit any conception of the Difference Engine as a thinking-machine as it is to disprove the claim that the chess-player is a pure machine. This he will attempt in a single move which both distinguishes and identifies the two machines in relation to one another and (on another plane) in relation to the human, and which pivots on the opposing concepts of determinism and contingency.

"What," Poe asks, "shall we think of the calculating machine of Mr. Babbage?" (Poe "Maelzel's" 319). The question is an interesting one: what shall we *humans* think about the image of a thinking-machine? Machines certainly are not included in the thinking "we" of Poe's query, and while excluding machines, the formulation draws attention to the thinking function of the human. Poe's "*a priori*" answer is that the thinking-machine cannot properly be thought. But this sets the stage for the next shift in Poe's essay, which becomes a tutorial in how to think, as "What shall we think?" becomes "*How* shall we think?"

Simply put, Poe tells us that we should think of the Difference Engine as a pure machine, which, as such, must function entirely deterministically, and "without the slightest intervention of the intellect of man" (Poe "Maelzel's" 319).[139] For Poe, this is a direct result of the Difference Engine's primary function in unerringly calculating algebraic tables:

> Arithmetical or algebraic calculations are, from their very nature, fixed and determinate. Certain *data* being given, certain results necessarily and inevitably follow. These results have dependence upon nothing, and are influenced by nothing but the *data* originally given. And the question to be solved proceeds . . . to its final determination, by a succession of unerring steps liable to no change, and subject to no modification" (Poe "Maelzel's" 319).

Because Babbage's Difference Engine embodies these deterministic rules,

> upon starting it in accordance with the *data* of the question to be solved, it should continue its movements regularly, progressively, and undeviatingly towards the required solution, since these movements, however complex, are never imagined to be otherwise than finite and determinate (Poe "Maelzel's" 319).

In other words, it is the machine's flawless embodiment of mathematical rule that renders its function deterministic, and, therefore, *conceivable*. That is, precisely because such a machine cannot be "imagined" (by Poe) as other than deterministic, it can, in fact, be imagined. "This being the case," Poe continues, and, we are to understand, *only if* this is the case, "we can without difficulty *conceive* the possibility of arranging such a mechanism" (Poe "Maelzel's" 319).

At the time Poe's essay was written, Babbage was already at work on designs for the Analytical Engine, which was conceived expressly in terms of *contingent* function. However, as Poe's main source for the Difference Engine seems to have been Willis/Brewster, and as the Analytical Engine (unlike the Difference Engine) was not widely publicized, Poe might be excused for his characterization of the Difference Engine as functioning entirely deterministically. But, Poe is not merely setting up a straw man, nor casually missing the greater significance of the Difference Engine. Rather, on one plane, he is using his "*a priori*" argument to conflate the two machines (the Difference Engine and the chess-player), and thus to correct regarding the Difference Engine what he sees as another iteration of essentially the same popular error as that made with regard to the chess-player. We have already seen the extent to which the press coverage in both Britain and the United States discursively constructed the Difference Engine as a thinking-machine (including Poe's main source, Brewster), largely influenced by Babbage's own characterizations of its contingent functioning. The force of Poe's argument regarding the Difference Engine is to discredit such claims, and the conception of the thinking-machine in general.

The discussion simultaneously serves as the *counter* example in Poe's application of his "*a priori*" argument to the case of the chess-player. Unlike the Difference Engine, the chess-player does not follow a determined course from the initial input of fixed data, for, in the game of chess

> [n]o one move necessarily follows upon any other. From no particular disposition of the men at one period of a game can we predicate their disposition at another period" (Poe "Maelzel's" 319).[140]

There is, in other words, no mathematical rule or linear progression upon which to order a game of chess.[141] Rather, "in proportion to the progress made in a game of chess, is the *uncertainty* of each ensuing move" and the progress of the game depends upon "the variable judgment of the players" (Poe "Maelzel's" 319). Moreover, were the chess-player a pure machine functioning deterministically, it "would be necessarily interrupted and disarranged by the indeterminate will of his [human] antagonist" (Poe "Maelzel's" 319).[142] For Poe, then, the fact that the chess-player is not "interrupted and disarranged" by the contingent play of its human opponent (and here Poe strays somewhat from his *a priori* argument into empirical observation)[143] *proves* that the chess-player is neither deterministic nor a pure machine, but is, rather, "regulated by mind" (Poe "Maelzel's" 319).

But, again, Poe is arguing *a priori*, that is, tautologically, and this concept of mind appears to do nothing more than to reproduce the Cartesian association among the concepts of the human, contingency, and thought, where each concept entails, the others as necessary components. Poe's "*a priori*" argument, then, is simply this: since only human minds function contingently, and all (pure) machines function only deterministically, then any apparently pure machine that functions contingently is not, in fact, a (pure) machine, but must be controlled by a human mind. It is this argument that Poe believes renders the chess-player inconceivable as a thinking-machine. The corollary, that a pure machine must function deterministically, also renders the Difference Engine inconceivable as a thinking-machine, once it has been granted that it is a pure machine--a point that Poe takes as given.

This apparently airtight argument, however, begins to produce some strange effects. On the one hand, Poe strenuously insists upon the Cartesian identification of contingency with mind, and mind (or thought) with the human, disallowing even the conception of a chess-playing machine--that is, of a thinking-machine. At the same time, it is this unique ability of the human mind to function contingently that makes it the superior player,[144] so that somehow the inconceivable chess-playing machine turns out to be merely a poor opponent.

To complicate the matter further, these same conceptual alignments that lead Poe to proclaim the chess-playing machine to be both inconceivable and a poor player, lead him also to suggest that such a machine would, at the same time, inevitably be the *superior* player. Indeed, Poe points to the fact that the automaton does not win every game as further proof that it is "regulated by mind--by some person" (Poe "Maelzel's" 323).[145] "Were the machine a pure machine," Poe writes, "it would always win. The [deterministic] *principle* being discovered. [...] a farther extension would enable it to *win all* games" (Poe "Maelzel's" 323).[146] That is, playing deterministically, and therefore flawlessly, the thinking-machine, once conceived, is inevitably the superior thinker.

Thus, Poe's argument seems to put into play simultaneously two opposing concepts of thought. The first of these incorporates contingency as its identifying component concept in association with the human, while the second incorporates the identifying component concept of determinism in association with the machine.[147] Moreover--and this is the source of the essay's great anxiety--it is the image of the machine functioning more fully and completely as a thinker--that is, surpassing the human at the very function which identifies the human in opposition to the machine--that emerges fully conceived from Poe's *a priori* argument against the conceivability of a thinking-machine.

Moreover, the deductive method of reasoning modeled by Poe himself in the same essay seems to align his concept of mathematically deterministic thought--the concept of thought with which he has already identified Babbage's Difference Engine and the "pure machine"--with the human mind. And, not only does Poe identify this "automatic" concept of thought with the human, but he offers it as the highest form of human thought.[148] Indeed, Poe's purpose for the essay seems largely didactic--aside from providing the solution to the mystery of Maelzel's chess-player, Poe intends to use the problem in order to demonstrate the proper conduct of reason, that is the ideal functioning of the rational human mind. And, he states his concept of this ideal thinking quite clearly. We have seen that Poe considers himself to have already provided "a mathematical demonstration, *a priori*"[149] regarding the inconceivability of a chess-playing machine in principle, and has told us that such "[a]rithmetical or algebraical calculations are, from their very nature, fixed and determinate" (Poe "Maelzel's" 319). As far as Poe is concerned, then, the question of Maelzel's chess-player is "fully decided" by the end of the essay's second page (Poe "Maelzel's" 323).

But, again, Poe's purpose is largely didactic, and he wishes "to convince, in especial, certain of our friends upon whom a train of suggestive reasoning will have more influence than the most positive *a priori* demonstration (Poe "Maelzel's" 323).[150] Nonetheless, Poe's method of reasoning regarding the details of this particular thinking-machine is also self-consciously and didactically deductive, that is, deterministic. In fact, Poe takes pains to identify and provide examples of faulty reasoning practice on the parts of others. Thus, in a survey of the existing literature (again, drawn primarily from Brewster), Poe identifies several "absurd [. . .], silly [and] bizarre explanations [. . .] followed by others equally bizarre" (Poe "Maelzel's" 321-322).

The fact that all three solutions singled out by Poe vary only in small details from his own suggests that what is silly, absurd, and bizarre about them has primarily to do with his judgment that they were formulated through "a course of reasoning exceedingly unphilosophical" (Poe "Maelzel's" 321). Indeed, in the case of the most plausible of the three,[151] Poe praises the conclusion while disparaging the method, acknowledging that the *"results"* [Poe's emphasis] "are, undoubtedly, in the main just" (Poe "Maelzel's" 321); it is, in fact, the method more than the result that interests Poe. Simply, he is concerned with promoting an image of thought whereby "Certain *data* being given, certain results necessarily and inevitably follow" (Poe "Maelzel's" 319).

Thus, Poe explains that his further analysis will be "deduced" from "*observations* taken during frequent visits to the exhibition of Maelzel" (Poe "Maelzel's" 323).[152] "[W]e will in the first place," he writes, "endeavor to show how [the chess-player's] operations are effected, and afterwards describe, as briefly as possible, the nature of the *observations* from which we have deduced our result" (Poe "Maelzel's" 322). Here Poe stresses the significance of his own "observations," which should be read as "data," and explains that he has "purposely avoided any allusion to the manner in which the partitions are shifted," since "it is performed out of the way of the spectators"--that is, there is no specific data concerning it (Poe "Maelzel's" 323). Thus, along with several inductive speculations regarding the chess-player's operation,[153] the remainder of the text does, indeed, catalog a number of reliable observations from second-hand accounts as well as (presumably) his own first-hand observations, and the "certain results [which] necessarily and inevitably follow" (Poe "Maelzel's" 319).

Moreover, Poe's discursive style from this point seems designed to produce a deductive effect. Thus, the (apparently) precisely phrased observation that the "the moves of the Turk are not made at regular intervals of time, but accommodate themselves to the moves of the antagonist" suggests its status as raw data. This data, then, is fed into the deductive machinery of Poe's "*a priori*" argument, in order to "prove [. . .] that the Automaton is not a *pure machine*" (Poe "Maelzel's" 323]. Similarly, Poe's observation that when the machine was moved in order to allow inspection of the interior, "certain portions of the mechanism changed their shape and position in a degree too great to be accounted for by the simple laws of perspective" produces the conclusion that mirrors are used to "multiply to the vision some few pieces of machinery within the trunk" (Poe "Maelzel's" 323-324). The flickering of a candle reflected against interior portions of the machine shows that the interior mechanism is designed "to be easily slipped, *en masse*, from its position. [. . .] when the man concealed within brings his body into an erect position upon the closing of the back door" (Poe "Maelzel's" 324). And, the fact that the automaton plays with its left hand, demonstrates the precise position of the man within the Turk (Poe "Maelzel's" 326).[154] These, and other, observations and their conclusions, along with precise measurements of the Turk's exterior cabinetry, provide the data from which Poe "deduces" his solution.[155]

Thus, the question "What shall we think?" has become transformed into "*How* shall we think?" and Poe's answer to this question again is, "mechanically." That is, Poe's image of the ideal operation of the identifying human function, thought, as multiply actualized through the essay's analytical method of ratiocination, turns out to reproduce his image of Babbage's Difference Engine and the "pure machine." In other words,

Poe's arguments against the conceivability of a thinking-machine are, themselves, thinking-machines, and identify the human as a thinking-machine.

Thus, like Descartes, Poe wishes to define the human against the machine through a concept of thought which is characterized by (or as) contingency and free will; but the concept of (human) thought actualized by Poe in his essay elaborating that view turns out to be the very image of the thinking-machine, functioning *automatically* from given inputs to produce *determined* outcomes. Furthermore, Poe's image of thought itself is identified by determinism as its central component concept. Thus, aside from the fundamental incoherence of Poe's conceptual apparatus, this automation of reason produces a certain dehumanizing effect with regard to human-defining thought.[156]

In this multiplication and confusion of the Cartesian concept of the human as a thinking thing, we can begin to see the motivation for Poe's anxiety, and that the absoluteness of his insistence upon the inconceivability of a thinking-machine suggests that something more than a technical question is at stake. It is possible that Poe's anxiety might be explained by Ambrose Bierce in "The Damned Thing," where he writes that "We so rely upon the orderly operation of familiar natural laws that any *seeming* suspension of them is noted as a menace to our safety, a warning of unthinkable calamity" (*Can Such Things* be 96; emphasis mine).[157] But, in fact, the source of Poe's anxiety is much more specific. For, as we have seen, once we can conceive (of) a thinking-machine, we can no longer clearly identify the human. Thus, where identity is ontology, the conception of the thinking-machine is precisely the death of the human.

But, there is more going on than that. For, Poe's essay does not simply fail to defend a Cartesian concept of the human against the machine. Rather, the essay folds and multiplies the relations of the identifying components of these concepts. It is as if the baker's transformation cut through multiple dimensions of Poe's essay simultaneously, distributing determinism as an identifying component across both his concept of the pure machine and his concept of human thought, while at the same time identifying the latter as uniquely contingent.

As we will see, it is finally in Melville, Mitchell, and Bierce that this anxiety, or sense of the uncanny, is expressed in Oedipal tropes, discursively linking the conception of the thinking-machine with the death, or potential death, of the human progenitor. But, again, the source of this expressed anxiety lies within the realm of philosophy and concepts; the concept of the thinking-machine robs a Cartesian concept of the human of its identifying characteristics, and thus of its concept, that is, of its ontology, its being, its identity. In other words, to the extent to which it is allowed to persist, the concept of the thinking-machine distributes the component concepts of the Cartesian human into new arrangements around new concepts, not only displacing the human as the organizing principle of thought and free will, but eliminating the identity of the category entirely. This loss of the human itself in the face of the thinking-machine, then, is the source of Poe's great anxiety, and the reason why the conceivability of the thinking-machine must be denied at all costs.

Literary Anxieties about Thinking-Machines
By the latter half of the nineteenth century, the trope of the thinking-machine had become widely disseminated in fiction, and the figure of the android commonplace (Franklin 132). And, while not all did so, a fair number of these stories addressed some anxiety over the ontological confusions that they iterated.[158] Three such stories, Herman Melville's "The Bell-Tower," Edward Page Mitchell's "The Ablest Man in the World," and Ambrose Bierce's "Moxon's Master" will be examined briefly here.

Each of these stories, as an iteration of the thinking-machine, both separates the concepts of human and machine, and folds them back into one another in a unique rearrangement of their component concepts. The emerging patterns from these multiple transformations include the refusal of those concepts to line up predictably between the human and the machine, and an emerging anxious uncertainty about how to tell the difference between these presumably certain ontological categories. Moreover, in a pattern that will be repeated throughout that century and well into the next, the stories express the resulting anxiety about the loss of the human as an identifying ontological category distinct from the machine, through a broad Oedipal imagery and the trope of the reversal between master and slave.

"The Bell-Tower"

While it is not unambiguously a story about a thinking-machine, Melville's "The Bell-Tower" is worth considering here nonetheless, partly for that very ambiguity, but also because at least one scholar considers it to be the first "fully formed" English language story about a humanlike automaton,[159] and as such it expresses a profound anxiety over the usurpation of human identity by a machine made in its image (Franklin 135). Moreover, while it certainly cannot be reduced to an allegory about humans and machines,[160] the story at one level pivots precisely on the point of confusion between the machine conceived as something essentially different from a human being and the machine conceived as something functionally indistinguishable.

Thus, as Melville variously cuts and folds the human and machine into one another, confusing the essential conceptual components between the two, we are repeatedly made uncertain as to the domino's ontological status, and are "stirred" by "all sorts of vague apprehensions" (Melville 143). And, as this effect is produced, we are also made increasingly uncertain about other oppositions--such as those between master and slave, or vitalism and materialism--as well. Indeed, Bannadonna's "law forbidding duplicates" may be read as the logical analysis of identity itself--that is, where there is no identifiable difference between two things, there can be said to be only one thing

(Melville 146). That is, we may read the story on this level as an especially sinister iteration of Leibniz's Law, or the principle of the identity of indiscernibles, such that where we cannot tell the difference between the human and the machine, there is only the machine.[161]

Thus, the uncanny aspect of the story derives largely from the "[u]ncertainty" regarding every aspect of the event, which "may, or may not, have" resulted from some mechanical failure, but primarily from uncertainty over the domino's ontological status, and its implications for that of Bannadonna and the human (Melville 148-149). For example, Bannadonna observes that when viewed from below, "the human figure [. . .] undergoes such a reduction in its apparent size, as to obliterate its intelligent features," such that it appears "automatic" (Melville 149). But, though the automaton is described as an "object" and presented as "an elaborate piece of sculpture, or statue," one observer "thought" he noticed that the mechanism "was not entirely rigid, but was, in a manner, pliant,"[162] while another "ventured the suspicion that it was but a living man" (Melville 142-143).[163]

Furthermore, in the presence of the automaton, city officials hear sounds that may or may not be the wind, while we are provoked to wonder whether it may not be some breath or soul of the figure that they hear (Melville 144). Later, the gathering crowd hears what sound like "screams and plainings, such as might have issued from some ghostly engine," an uncanny suggestion of an animate and (therefore) tortured soul within mechanism (Melville 47). And, when the magistrates hear in the belfry what might be a "footfall," they are reassured by Bannadonna that "no *soul*" remains there. But, when he continues that "it better knew its place" in the Excellenza's presence, we are no longer certain as to what "it" refers, while at the same time, we are confronted by the disturbing image of a slave who might *not* know his place (Melville 146). Even Bannadonna seems uncertain as to his creation's ontological status, stammering over a reference to "him? *it*," while revealing the automaton's name. "Haman," of course, looks and sounds almost "human," while simultaneously invoking the threat of racial extermination (Melville 144).[164]

But, Bannadonna's own status is ambiguous as well. Thus, for instance, we are told repeatedly of his uncertain parentage (Melville 140, 144). Described as a "true artist," he valorizes the contingent differences between each iteration of his art, desiring his automaton to be "an original production" (Melville 146, 152). But, Bannadonna is also "the mechanician," a "practical materialist," conquered in the end by the deterministic principles that he has served, and by the creature that turns out to be his double after all (Melville 151-152). Moreover, Bannadonna would seem to have lost his soul much earlier, and thus ultimately himself, in the catastrophic murder of the workman, for it is through this act that the tower, and Bannadonna's ambitions, at last fail.

While in many ways invoking the homunculus of an older magic,[165] Melville's domino[166] is clearly recognizable as a mechanical construction, the masterpiece of "the great mechanician," an automaton functioning as an integral component of Bannadonna's elaborate clockwork mechanism (Melville 140). Indeed, the domino is an example of precisely the kind of clockwork figures that had so captivated Descartes in his youth, and later informed Babbage's calculating engines. As both a homunculus and as an automaton, then, Melville's domino presents the uncanny image of a human body without a human soul, iterated within a self-conscious reference to Cartesianist ontology and contemporary technologies. And, like Babbage's Difference Engine, Bannadonna's automaton is conceived not only as his double, but as his superior, as he resolves to "devise some metallic agent, which should strike the hour with its mechanic hand, with even greater precision" than the human could achieve (Melville 149).

Thus, once wound, the mechanism follows Bannadonna's instructions "precisely" and "infallibly"--that is, mindlessly-- and yet at the same time unpredictably, against the interests and wishes of its master--*as if* it had a mind of its own. Describing the bell-tower's anticipated first ringing of the hour, Bannadonna predicts that "that stroke shall fall there, where the hand of Una clasps Dua's. The stroke of one shall sever that loved clasp" (Melville 144). And, indeed, it is the stroke of one, that is of Bannadonna's automaton, which severs the former's grasp upon his double. But, this act also has the effect of severing the doubling between soul and body, between master and slave, between human and machine, killing the human master, and leaving only Bannadonna's "manacled" and "iron slave," the (perhaps) thinking-machine still standing (Melville 144).

Once the uncanny ambiguity of its ontological status has been discovered, Bannadonna's domino, like Descartes' daughter, is shrouded and sunk deep into the sea (Melville 149). But, we are left nonetheless with the persistently uncanny-- that is, ontologically ambiguous--image of the "blind slave" standing over the body of the dead master, creator, father, still "whispering some post-mortem terror" (Melville 152).

"The Ablest Man in the World"[167]
Edward Page Mitchell's 1879 short story "The Ablest Man in the World" plays upon this same terror, but locates it ironically through the character of Mr. Fisher, a comically (or perhaps tragically) patriotic, and profoundly unscientific, American.[168] The story begins with the American's accidental discovery, while traveling in Europe, that the internationally renowned and brilliantly accomplished Baron Savitch, by all accounts "a blessed man," is, in fact, a cyborg.

That there is something strange or uncanny about the Baron is suggested even before Fisher encounters him, as he hears reports of some "terrible fit" that has rendered a valet "beside himself with terror" and the Baron himself "desolate with apprehension" (Mitchell 26-27). This impression is reinforced by the Baron's painful contortions, which "distort the natural expression of his face" (Mitchell 27). Moreover, we are told that there is a "mystery about his origin that had never been satisfactorily solved," and that "no one knew with positive certainty his father's name" (Mitchell 31-32). Rapperschwyll's unexpected return to Switzerland and his own mother's deathbed toward the end of the story reinforces the "unnatural" or uncertain origins of the Baron. Thus, even before we learn anything specific about the Baron's condition, we have already been prepared to find the details surrounding him to be unnatural, mysterious, and terrifying.

But, none of this prepares Fisher, or presumably, the reader, for the disturbing emotional impact of his discovery of the cyborg's "confounded secret" (Mitchell 30). Confronted and blackmailed by Fisher, the Swiss watchmaker Rapperschwyll[169] recounts the truth of the Baron's mysterious origins in some detail--but in so doing raises deeper uncertainties about Savitch's ontological status and the meaning of his origins (Mitchell 33).

For, unlike virtually every cyborg to appear since, Mitchell's Baron Savitch is the union of a human body and a mechanical brain.[170] This body, belonging to the unfortunate Stépan Borovitch, "a boy of eleven [who] [s]ince he was born had not seen, heard, spoken, or thought," thus appears, by the doctor's Cartesian logic, to harbor no soul (Mitchell 38). In other words, without speech, and without reason, the boy is in Cartesian terms the ontological equivalent of an animal, which is, of course, to say a machine. It is for this reason that, Dr. Rapperschwyll believes that he can, without moral objection, "operate on that poor, worthless, useless, hopeless travesty of humanity as fearlessly and as recklessly as upon a dog bought or caught for vivisection," and install within that soulless body the heir of "Babbage's cogs and pinions" (Mitchell 38, 36). Thus, Rapperschwyll, in revealing the cyborg's origins, also suggests the full significance of the Baron Savage's name, and points toward the uncertain ontology that it signifies: as a mechanism, he lacks the "vital principle" that would involve him in a natural process of human reproduction; while as a mute, blind, and deaf presumed idiot, he lacks the higher functions of thought that have uniquely identified the human since Descartes (Mitchell 37). In other words, despite appearances, the "barren savage" is not human.

But, despite Rapperschwyll's assurances, we are left with lingering questions about Stépan Borovitch's ontological status. Indeed, it is Rapperschwyll himself who tells Fisher that Stépan had often enjoyed sitting in the sun, murmuring and mumbling in satisfaction, ambiguous acts that might or might not have indicated higher functions of thought. Most disturbingly, Rapperschwyll seems to have suspected the boy's humanity, explaining that nature "had *walled in his soul* most effectively," a very different thing than having no soul at all (Mitchell 38; emphasis mine). Thus, when the boy's brain is removed and replaced with a fully functioning mechanical one, we are uncertain as to whether it is the idiot Stépan, the cyborg Baron Savitch, or the materialist Dr. Rapperschwyll who represents the true monster.

We are certain, however, that the Baron Savitch's mechanical brain is a thinking-machine "far beyond Babbage's in its powers of calculation" (Mitchell 36). Indeed, as Rapperschwyll insists, this "new machine was fed with facts, and produced conclusions. In short, it *reasoned*" (Mitchell 36). And, in this strange way, it seems to be the mechanical brain which gives Borovitch/Savitch his humanity, that is his speech, his reason, his *soul*. In other words, in a complete confusion of Cartesian concepts, this human body appears to harbor a *mechanical soul*, and the Baron Savitch appears to be an "impossible person" (Mitchell 40). But, this configuration involves a second transformation not unlike that effected by Poe, and it is this second transformation that keeps Fisher (as well as, perhaps, the reader) "continually in a state of distraction" (Mitchell 40).

The title itself ironically suggests this anxiety. For, if the ablest man in the world turns out to be a machine, then it is no longer clear that the machine is not in fact human, that there are any uniquely human functions at all, or even that there is any such thing as a human. The concept appears to be empty. Moreover, the machine's superiority to the human, precisely with regard to those functions which identify the human, renders the mastery of the human creator over his machine creation precarious at best.

For, the brain of Baron Savitch is not *merely* human in its reasoning abilities; it is, rather "superhuman" (Mitchell 32). Thus, Savitch is already well-known internationally for his "cool, unerring judgment, [and] far-reaching sagacity" which "never errs; for the machine that reasons beneath his silver skull never makes a mistake" (Mitchell 32, 36). By contrast, we are reminded, "the results of human reasoning are often, if not always, false" (Mitchell 39). And, thus, reason itself appears to be a "mechanical" function "unalterable as the laws of arithmetic," that is, thoroughly deterministic, while the human continues to be identified as and with contingency expressed as error and emotion (Mitchell 36). In this way, reason has crossed the human/machine boundary to become an identifying component of the machine, while the machine has become "the ablest man in the world" (Mitchell 39).

In this way, the machine does not simply outperform the human, but materially and conceptually replaces it. Thus, the "artificial intelligence that operates with the certainty of universal laws" will overcome "his inferiors who reason falsely," Rapperschwyll explains; *"the ultimate evolution of the creature is into the creator,"* and it is the creature's superior ability which will make him, inevitably, "master of the world" (Mitchell 38-39).

But, Mitchell's narrative complicates rather than simply reproducing these anxieties, posing serious questions about what it means to be human, and whether we can in fact be human once we have conceived a thinking-machine. Thus, despite Rapperschwyll's ambitions and Fisher's "unspeakable consternation," the cyborg itself appears to be completely unmotivated by dreams of either mastering or destroying his human creators (Mitchell 40). In fact, there is an innocence and sweetness to the Baron's demeanor that even Fisher feels. And, contrary to Rapperschwyll's dissertations upon the unerring mathematics of mechanism, it is the Baron's "high-strung mental organization [that] render[s] him susceptible to sudden and alarming attacks of illness" (Mitchell 32). Moreover, it is precisely the Baron's romantic attentions toward Miss Ward, a young woman in Fisher's party--that is, the "transient whim of an automaton"--that finally incites Fisher to action (Mitchell 41). Thus, it is not only the cyborg's ontological status as human or machine that is uncertain, but also which of these would pose the bigger threat.

But, complicating matters further, this uncertainty over Savitch's ontological status runs parallel to confusions about Fisher's identity as well. Thus, we are told that he is randomly and routinely (mis)identified as "Herr Doctor Professor" upon his arrival in Baden (Mitchell 25). And, it is this mistake that directly involves him with Savitch, as he is called upon to tend to the Baron's ill health in the temporary absence of Rapperschwyll. In this way, Fisher is figured as the double of Rapperschwyll, later exhibiting the same "ruthlessness" toward Baron Savitch that Rapperschwyll had demonstrated with regard to Stépan Borovitch. More obviously, Fisher's "imperturbable coolness" in dealing with Rapperschwyll marks him as a double for Savitch, who is repeatedly, if erroneously, described as "cool" or "cold" (Mitchell 30, 32, 33).

The effect of all of these uncertainties,[171] however, is to shift the location of the reader's anxiety, through Fisher's, away from ontological issues regarding the mechanical threat to the human, and to relocate it as anxiety *about* that anxiety. "Heaven forgive me if I am making a fearful mistake!" Fisher thinks to himself as he stands on the verge of destroying the suddenly "wondrous machine," and "put[s] his hands to his ears to shut out the sound" of the "wild, despairing cry" which only "may have been the gull's" (Mitchell 43-44). But, what kind of mistake could it be to destroy a *machine*? What crisis of conscience could the act of destroying a *mere* machine provoke? As the story ends, we are forced to wonder, with Fisher, whether something more sinister has not taken place, whether Fisher himself has not destroyed something human, or as good as human, as he reports that he has "throw[n] overboard the Baron Savitch" (Mitchell 44).

"Moxon's Master"
Appearing in his 1893 short stories collection, Bierce's "Moxon's Master" provides what appears to be a direct answer to Poe and the question that he raises about the automaton chess-player, namely, *Can Such Things Be?*[172] Like Poe's essay, Bierce's story includes a detailed description of the chess-player and a lengthy dissertation upon the conceivability of the thinking-machine. In contrast, however, to the reactive anxieties of Poe and Melville, or even to the thoughtful questioning of Mitchell, Bierce seems eager to expose, indeed to embrace, the ontological uncertainties that such questions pose.[173]

Indeed, the ontological transformations effected by questions about the conceivability of the thinking-machine are exposed through the text's opening dialogue, which answers the narrator's query, "do you really believe that a machine thinks?" with Moxon's response, "What is a 'machine'? [. . .] is not a man a machine?" (Bierce "Moxon's Master" 27). Thus, we are immediately confronted, by this "inventor and constructor of machines," with the proposition that questions about the boundary between the human and the machine involve a radical challenge to the ontological status of the categories themselves (Bierce "Moxon's Master" 30).

Moreover, it is not certain either that thought itself has any special metaphysical or conceptual status, except as a description of the contingent functioning of goal-seeking, that is, self-organizing, systems. For while a man, perhaps, only "thinks he thinks," the abilities of plants to act and react to changes in their environments, as well as the crystallization of minerals, "proves that they think" (Bierce "Moxon's Master" 27-28). The opposition between vitalism and materialism is similarly transformed, as "life" also becomes the functional "force" of "intelligence and purpose" which again names the goal-seeking function of self-organizing systems, and which is expressed by matter "in proportion to the complexity of the resulting machine and of its work" (Bierce "Moxon's Master" 29).

Thus, it is not surprising that the narrator at first mistakes Moxon's mechanical chess-player for "another person" even though Bierce's readers would have already recognized the automaton chess-player, which the narrator describes using familiar details, including a "crimson fez" (Bierce "Moxon's Master" 31-32). But, even this identity is not certain, for, unlike Maelzel's chess-player, Moxon's automaton holds his left arm on his lap, plays with his right, and bears more resemblance to a gorilla than to a man (Bierce "Moxon's Master" 30).[174]

But the ontological uncertainties surrounding the chess-player are more serious, as it seems to exhibit both the contingent functioning that identifies the human and the deterministic movements associated with the machine. For example, the narrator observes the automaton's play as "slow, uniform, mechanical and, [. . .] somewhat theatrical" (Bierce "Moxon's Master" 32). More dramatically, upon losing its game, the automaton shrugs in an "entirely human gesture" that precedes what can only be described as a very human temper tantrum. A "disordered mechanism," the machine has "escaped the repressive and regulating action of some controlling part," and thus also of its creator and master, Moxon (Bierce "Moxon's Master" 33).

Thus, the narrator's answer to Moxon's opening question suggests something about what is at stake in maintaining the (illusion of) this boundary, for, by "machine," the narrator tells Moxon, he means "something that man has made and controls" (Bierce "Moxon's Master" 27). It is the latter point to which the title ironically refers, and upon which the drama and anxiety of the text plays. It is, indeed, precisely at the moment in which the narrator first sees the chess-player, though he mistakes his identity, that he is gripped with a feeling that he is "in the presence of an imminent tragedy" (Bierce "Moxon's

Master" 32). And, it is with the uncanny image of this tragedy that Bierce leaves us, Moxon on the floor,

> his head still in the clutch of those iron hands [. . .] his eyes protruding, his mouth wide open and his tongue thrust out; and--horrible contrast--upon the painted face of his assassin an expression of tranquil and profound thought (Bierce "Moxon's Master" 33).

Thus, it is precisely this mislocation of contingency within mechanism, and the ontological uncertainty that it produces, which creates the story's uncanny effect.[175] In other words, it is the chess-player's contingent functioning that makes him a sore loser, and the chess-player both loses and becomes angry about losing precisely because he is *not* a deterministically functioning automaton. Or, to trace the confusion another way, as a thinking-machine, the chess-player functions contingently, which is to say imperfectly and emotionally, which is to say irrationally. In other words, Bierce's thinking-machine lacks reason, and thus effects a complete (and delighted) perversion of the Cartesian categories defended by Poe--as well as an exposure of the incoherence of his defense.

Thus, each of these three texts, as an iteration of the thinking-machine, effects a transformation in which the component concepts of the human and the machine are reshuffled and rearranged in multiple configurations which destabilize the ontological status of both human and machine. But, at the same time, these literary "mash-ups" begin to produce recognizable patterns of anxiety.

Thus, despite Freud's assertion that "intellectual uncertainty is [. . .] incapable of explaining" it, Ernst Jentsch seems to have been correct when he suggested that it *is* uncertainty, or "doubt as to whether an apparently living being is animate and, conversely, doubt as to whether a lifeless object may not in fact be animate" that produces anxiety, or a feeling of uncanniness, particularly with regard to automata exhibiting "certain bodily or mental functions" (Freud 938; Jentsch 11,

13). But, as Jentsch failed to recognize, this is precisely because such uncertainties cast doubt upon the *ontological* status of the categories themselves. In other words, it is through the Oedipal narrative that anxieties over category confusion and the loss of (human) identity are expressed, not, as Freud suggests, the reverse.

CHAPTER FOUR/ FOURTH ITERATION: TRANSFORMATIONS

> "We love a correct definition. The Automaton Chessplayer was either a gross piece of humbug, or it was a sentient being, endowed, like man himself, with volition, judgment, and all the rest of it; but in neither case was it an automaton" (Walker 5).

Contrary to the predictions of Descartes, the 1940s and 1950s saw a proliferation of electro-mechanical devices which seemed to reproduce many functions of human thought. These material iterations of the thinking-machine, popularly figured as "mechanical brains" (Berkeley 1), included military targeting systems, code-breaking machines, the first modern programmable computers, and (despite Poe) actual chess-playing machines. But, these new images of the thinking-machine also called into radical question the ontological status of both the human and the machine, transforming and destabilizing the Cartesian concepts grounding the popular question which they provoked: "Can machines think?" (Berkeley 1). Indeed, it is through the multiplicity of attempts to answer this question that the trope of the thinking-machine itself is multiplied and transformed, ultimately encompassing those contingent functions which had identified a uniquely human ontology within a Cartesian logic. Moreover, as the component concepts of the human (e.g. contingency and thought) were transformed and redistributed over a broader ontological field, the trope of the human itself became an empty signifier, largely falling out of these discourses.

Wiener and Communicating Machines

In the late winter of 1943-1944, Norbert Wiener and Max von Neumann organized the first "joint meeting of all those interested in what we would now call cybernetics. [. . .] Engineers, physiologists, and mathematicians were all represented" (Wiener *Cybernetics* 14). Frustrated by the lack of communication among researchers from different fields working on the same technological and theoretical concerns, Wiener and his research partner Arturo Rosenblueth

> had already become aware of the essential unity of the set of problems centering about communication, control, and statistical mechanics, whether in the machine or in living tissue [but felt] seriously hampered by the lack of unity of the literature concerning these problems, and by the absence of any common terminology, or even of a single name for the field (Wiener *Cybernetics* 11).

For Wiener and Rosenblueth, words and concepts were precisely the information technologies through which the science and philosophy "of every age" were organized (Wiener *Cybernetics* 38). Thus, the absence of an adequate "vocabulary," was also precisely the absence of a conceptual schema through which to conceive the new science. In other words, neither the inherited disciplinary paradigms, nor the Cartesian concepts which grounded them,[176] would be adequate to the task of conceptualizing the new theoretical and material technologies which organized these mathematicians, biologists, and computer engineers around a common set of problems.   Moreover, as communication was already one of its central tropes, from a cybernetics point of view, the conscious development of this new discursive field was precisely the emergent organization of the field of cybernetics itself. Thus, by the end of the Princeton meeting, "it had become clear to all

[. . .] that some attempt should be made to achieve a common vocabulary" (Wiener *Cybernetics* 15).

Indeed, that the vocabulary, or concepts, as well as the disciplinary boundaries, of his own age were in the process of a profound transformation seems to have interested Wiener as much as did the technological achievements to which he contributed, and, thus, he consistently draws attention to the "boundary regions" (Wiener *Cybernetics* 2) between different disciplines and oppositional concepts.[177] But, in casting this organizational problem as one of communication, Wiener makes the radical suggestion that, without the emergence of a new discursive field, he and his colleagues literally could not conceive the thinking-machines that they were materially creating. Thus, Wiener insists that

> such words as life, purpose, and soul are grossly inadequate to precise scientific thinking. These terms have gained their significance through our recognition of the unity of a certain group of phenomena, and do not in fact furnish us with any adequate basis to characterize this unity" (*Human Use* 31).

Thus, functioning as a nexus where systems converge and emerge, Wiener participates in the process of organizing both the conceptual and the material technologies of thinking-machines as iterations of the posthuman.

Wiener was born in the United States in 1894, and as a self-described child prodigy, studied mathematics at Tufts, zoology at Harvard, and philosophy at Cornell, earning a Ph.D. in mathematics from Harvard at the age of eighteen. Wiener then spent the next few years teaching and studying mathematics, logic, and philosophy, including additional post-graduate studies under Bertrand Russell. After World War II, he settled at MIT as a professor of mathematics, where he remained. But it was Wiener's work on ballistics systems during World War I, and on targeting mechanisms for anti-aircraft guns during World War II, that began to focus his interests around problems of what he would eventually come to call "communication and control"--or, simply, "cybernetics" (Wiener *Cybernetics*; Wiener *Ex-Prodigy*; Wiener *I Am a Mathematician, passim*).

Specifically, Wiener helped to develop targeting systems for anti-aircraft artillery which could anticipate and predict trajectories, learning and adapting to the contingent evasive maneuvers of their human targets in order to "shoot the missile [. . .] in such a way that missile and target come together in space at some time in the future" (Wiener *Cybernetics* 5).[178] "[P]redicting the future" (Wiener *Cybernetics* 5), however, was the central problem, since Heisenberg's Uncertainty Principle, and quantum mechanics generally, made the deterministic prediction of any future event theoretically impossible-- particularly when one was specifically interested in predicting an object's precise location at a given point in time (Wiener *Cybernetics* 9).[179] That is, because the human-aircraft target was not a completely determined system, but an open system capable of decision and contingency, simple linear equations could not predict its path.

Wiener's way around this theoretical limit involved reconceptualizing, and expanding the scope of the problem, using data about the limits and capabilities of the airplanes, experience about the most common evasive moves against a given situation, and even data from past experience with specific human pilots, in order to develop non-linear equations which would predict a "cloud" of statistical probability for the airplane's location at a given time. The goal, was to mechanize the processing of experience about pilot-plane systems, and, ultimately to build in feedback mechanisms which would allow the device to do this learning from experience for itself. Thus, Wiener writes that,

> [t]he predictor uses the immediate past of the flight of the airplane as a tool for the prediction of the future by means of a linear operation; but the determination of the correct linear operation is a statistical problem in which the long past of the flight and the past of many similar flights are

used to give the basis of the statistics (*Cybernetics* 173). These "long-past statistical algorithms" are necessarily "highly non-linear," such that one may say that, "[i]n general, a learning machine operates by non-linear feedback" (*Cybernetics* 173).[180]

In other words, while the basic operations of the targeting device could be described determinately, and could be preprogrammed into the machine by its human engineers, the machine's ability to adapt to real-life contingencies required a different kind of programming--the functioning of which would itself never be wholly predictable by its designers.

What struck Wiener most forcefully about this work on targeting systems, was their similarity to functions of the human central nervous system (including the brain):

> it became clear to us that the ultra-rapid computing machine, depending as it does on consecutive switching devices, must represent almost an ideal model of the problems arising in the nervous system. The all-or-none character of the discharge of the neurons is precisely analogous to the single choice made in determining a digit on the binary scale" (Wiener *Cybernetics* 14).[181]

Thus, for Wiener, just as there is no clear boundary between the human pilot and the airplane,[182] there is also no clear boundary between the human body and the brain, nor between the brain and the mind.

In this way, questions about ontological boundaries are transformed, becoming questions about function instead--and specifically, questions about how dynamic systems respond contingently to, and organize themselves in response to, outside interference. Thus, Wiener and Bigelow concluded that

> the problems of control engineering and of communication engineering were inseparable, and that they centered not around the techniques of electrical engineering but around the much more fundamental notion of the message, whether this should be transmitted by electrical, mechanical, or nervous means (Wiener *Cybernetics* 8).

In other words, for Wiener, it is the functional organization of matter--not its ontological status as substance--that has meaning.

To some degree, then, Wiener's attempts to answer questions about the ontological status of machines involve demonstrations of their functionality as living beings--that is transformations of the concepts of life, the human, and mechanism. For instance, he writes quite uncontroversially in 1961 that, "phenomena which we consider to be characteristic of living systems are the power to learn and the power to reproduce" (*Cybernetics* 169). However, he goes on to say that we ought to read these separate functions as variants of the ability to become something new in response to changes in the environment. Thus, an "animal that multiplies is able to create other animals in its own likeness [but] not so completely that they cannot vary in the course of time" and an "animal that learns is one which is capable of being transformed by its past environment into a different being" (*Cybernetics* 169). Significantly, the functional distinction that Wiener makes in this passage between reproduction and production is precisely the difference between determinacy and contingency.

In cybernetic terms, communication, (that is, the deterministic reproduction of information) describes the process by which a system tends to organize another system as a replica of itself, while (the system's openness to outside) interference functions as the contingency which both disrupts the precise reproduction of the system, and enables learning, evolution, and thought (that is, *becoming*). In other words, "just as entropy is a measure of disorder, so information is a measure of order" (Wiener *Human Use* 116), and the self-organizing system emerges from the play between these two tendencies. Thus, this other side of communication and control, interference or contingency, plays a crucial role in the adaptability of self-organizing or complex (that is, intelligent) systems.

Thus, the concept of life becomes secondary to its

capacity to function contingently, and the details of its materiality are important only to the degree that they enable and organize that functionality. Thus, "to be alive" is not, for Wiener, a matter of being composed of any particular substance, but rather, "is to participate in a continuous stream of influences[183] from the outer world, in which we are merely the transitional stage" (Wiener *Human Use* 122).

In this way, Wiener redefines our concept of a "living organism" as that which engages in "[b]oth ontogenetic and phylogenetic learning," and our concept of the "human" as that living organism whose phylogenetic learning is primarily geared toward "establishing the possibility of good ontogenetic learning" (Wiener *Cybernetics* 169-170).[184] In other words, a human is a thinking-thing not because it is the substance of thought, but simply because it does in fact think. Thus, the question, "Can machines think?" is also transformed for Wiener, becoming not a question of ontology, but rather a more interesting set of questions about whether non-human machines can respond to contingency sufficiently to learn either ontogenetically or phylogenetically (Wiener *Cybernetics* 170).

Wiener's answer to the first part of the question draws upon his work with targeting systems during World War II. This, as we have seen, took advantage of the Heisenberg Uncertainty Principle, using non-linear (nondeterministic) second-order statistical models combined with real-world experience to make decisions about which first-order linear (deterministic) programs to execute (Wiener *Cybernetics* 5-11; 173). Comparing this relatively simple learning machine, as well as a hypothetical chess-playing machine, variously to a fight between a mongoose and a cobra, a bull fight, a sword fight, and a tennis game, Wiener concludes that these "physical contests and the sort of games which we have supposed the game-playing machine to play both have the same element of learning in terms of experience of the opponent's habits as well as one's own" (Wiener *Cybernetics* 172-175).

In other words, thought is conceptualized as a function

of learning, and, as we have seen, learning, for Wiener, is the process by which self-organizing systems transform themselves, or *become*, in relation to interference from their environment. Thus, questions about the boundaries between machines and humans, or mechanism and life, become for Wiener nonsensical, as the questions and the concepts which ground them are replaced with questions about contingent function.

Wiener similarly transforms the question of the "reproductive" capacity of a machine, which he describes as "not only a form of matter, but an agency for accomplishing certain definite purposes," while "self-propagation is not merely the creation of a tangible replica [but] the creation of a replica capable of the same *functions*" (*Cybernetics* 177-178; emphasis mine). Here, Wiener first shows how a "machine" may be said to function in a manner that is normally identified with "life," describing how a white box can be made to "potentially assume the characteristics of any non-linear transducer whatever, and then to draw it into the similitude of a given blackbox transducer" by a process of random interference, feedback looping, and the averaging of difference (*Cybernetics* 180).[185] In this way, one mechanical system may "capture" another mechanical system into its pattern, pulling it toward the same function, in effect mimicking the reproductive processes of life.

But in a second transformation, Wiener figures DNA as something like a reproducing machine, concluding that the transducer's behavior is "philosophically very similar" to

> what is done when a gene acts as a template to form other molecules of the same gene from an indeterminate mixture of amino and nucleic acids, or when a virus guides into its own form other molecules of the same virus out of the tissues and juices of its host (*Cybernetics* 180).

In other words, the machine is analogous to the organism, and the organism is analogous to the machine. But, it is the *function* of self-organization as a series of decisions unfolding through time that interests Wiener, and questions about ontological boundaries are dismissed as being either nonsensical or beside the point. Explaining the shift from Newtonian to quantum physics, Wiener stresses the forward unidirectionality of time such that, "[i]n tidal evolution as well as in the origin of species, we have a mechanism by means of which a fortuitous variability [...] is converted into a dynamical process into a pattern of development that reads in one direction," that is, *becoming* (*Cybernetics* 36). Thus, "the problem as to whether the machine is alive or not is, for our purposes, semantic and we are at liberty to answer it one way or the other as best suits our convenience" (Wiener *Human Use* 32).

It is clear that Wiener delights in discursively disrupting the normal alignments between inherited Cartesian concepts of the human (or life) and the machine, and the functions supposedly essential to each. But, it is also clear that this Cartesian language is, in fact, not adequate to clearly describe the real science and technology emerging from his work. Thus, Wiener is repeatedly in the position of making apparently nonsensical statements about both humans and machines. And this nonsense--or interference in the system of Cartesian discourse--is an important aspect of his rhetoric, and of his multiple transformations, as he typically shuffles and juxtaposes incompatible substances and functions, or discursively conflates opposing Cartesian categories under cybernetic self-organization.

This linguistic play is evident, for instance, when Wiener writes that, "the computing machine, and consequently the brain, is a logical machine. It is by no means trivial to consider the light cast on logic by such machines, both natural and artificial" (*Cybernetics* 125); and similarly, when he writes that, "in the nervous computing machine it is highly probable that information is stored largely as changes in the permeability of the synapses, and it is perfectly possible to construct artificial machines where information is stored in that way" (*Cybernetics* 130), or that "the machine may also be a communicative organism" (*Human Use* 136). Other examples of this word play include Wiener's suggestions that, "a large computing machine, whether in the form of mechanical or electric apparatus or in the form of the brain itself" (*Cybernetics* 132), "must be a logical machine and must combine contingencies in accordance with a systemic algorithm" (*Cybernetics* 118), and that "those living machines which we call animals" (*Cybernetics* xv) "are not living below the molecular level" (*God and Golem* 46).

Again, Wiener's use of language produces multiple images of the thinking-machine, exposing the inadequacy of the concepts grounding questions about the boundaries between human and machine. Thus, rejecting both the dualism and the materialism grounded in popular Cartesianism, he insists that "[i]nformation is information, not matter or energy. No materialism which does not admit this can survive at the present day" (*Cybernetics* 132). Thus, because "many automata of the present age are coupled to the outside world both for the reception of impression and for the performance of actions [. . .] they can be subsumed under one theory with the mechanisms of physiology" (*Cybernetics* 43), while the new physiology constitutes the "study of automata, whether in the metal or in the flesh [and] is a branch of communication engineering" (*Cybernetics* 42).

Furthermore, these "automata," no longer automatic at all, embody the contingent functional analogs of both memory and learning (*Cybernetics* 43). Thus

> the modern automaton exists in the same sort of Bergsonian time as the living organism; and hence there is no reason in Bergson's considerations why the essential mode of functioning of the living organism should not be the same as that of the automaton of this type. Vitalism has won to the extent that even mechanisms correspond to the time-structure of vitalism; but, as we have said, this victory is a complete defeat, for from every point of view which has the slightest relation to morality or religion, the new mechanics is as mechanistic as the old. Whether we should call the new point of view materialistic is largely a

98 | DESCARTES' DAUGHTERS

question of words [...] In fact, the whole mechanist-vitalist controversy has been relegated to the limbo of badly posed questions" (*Cybernetics* 44).
Cybernetics, then, is the new set of questions that emerges from "the central point in the old controversy between vitalism and mechanism" (*Cybernetics* 38)--and between the human and the automaton.

Wiener's Chess-Players
Despite an aversion to the ontological opposition between humans and automatons, Wiener wrote a great deal about "the question of whether it is possible to construct a chess-playing machine, and whether this sort of ability represents an essential difference between the potentialities of the machine and the [human] mind" (*Cybernetics* 164). His earliest treatment of the subject occurs in *Cybernetics*, as a final note to the final chapter, somehow functioning both as an afterthought and as the culminating question to which the entire exercise has led. Casting aside the extremes of a machine which would "play an optimum game in the sense of von Neumann," or a machine which would "play in the sense of following the rules, irrespective of the merit of play," Wiener considers "possible" a machine which would "offer interesting opposition to a [human] player" of average skill level. And, in a fair reproduction of Babbage's proof for mechanized chess, Wiener imagines that such a machine would evaluate its own moves in light of the contingencies of its opponent's play, looking "two or three moves ahead," (165).[186]

Thus, in an interesting transformation of Poe's logic, Wiener concludes that such a machine
> might very well be as good a player as the vast majority of the human race. This does not mean that it would reach the degree of proficiency of Maelzel's fraudulent machine, but, for all that, it may attain a pretty fair level of accomplishment (Wiener *Cybernetics* 165).

That is, its deterministic programming, far from making the

automatic chess-player a perfect player, as Poe suggested, would limit its ability to predict its opponent's moves far into the future, and thus render it a merely average player.

In *The Human Use of Human Beings*, Wiener acknowledges that "the question has a considerable history behind it," recalling that "Poe discussed a fraudulent machine due to Maelzel, and exposed it; showing that it worked by a legless cripple inside" (Wiener *Human Use* 175).[187] His discussion here recounts some of the work of von Neumann and Shannon toward solving the problem, focusing primarily upon one of Poe's main objections, namely that it would be impossible to program a computer in advance for every contingency of the game. Echoing the passage from *Cybernetics*, Wiener counters this objection by suggesting a weighing system which could guarantee, not a perfect game, but, "the best that can be done for a limited number of moves ahead" (*Human Use* 175).

Among the material added to the 1961 edition of *Cybernetics*, however, is a more thoughtful discussion of the problem of chess-playing machines, one which connects insights based on the non-linear feedback and communication mechanisms of Wiener's early work on aiming systems, theories of learning as self-organization, and game theory. A similar discussion appears in *God and Golem*.

In these later discussions, the (by this time) fairly standard description of a chess program capable of weighing possible moves as far as three moves ahead, based on conventional valuations, in order to determine "its definitive play" (*Cybernetics* 165), is the starting point rather than the goal (*Cybernetics* 172). The limit of this sort of machine, Wiener points out, is that the human opponent can learn its "chess personality," and, after a few games, easily predict its behavior such that "if any trick [. . .] will work, then it will always work under the same condition" (*Cybernetics* 172). But, Wiener does not consider this kind of deterministic functioning to be an *essential* limit of the digital computer. Nor is such automatic play *limited to* the computer. On the contrary, in *God and Golem*,

Wiener allows this rigid sort of chess-player to be either human or mechanical--either can be "a mechanized player who does not learn," and "the player, be he a man or a machine, who plays by a simple table of merit [...] will give the impression of a rigid chess personality" (20).

In contrast, Wiener imagines a more flexible program, such that, while the chess-player begins with a determined set of rules for choosing moves, it also, at intervals,

> examines all the previous games which it has recorded on its memory to determine what weighting of the different evaluations of the worth of pieces, command, mobility, and the like will conduce most to winning. In this way, it learns not only from its own failures but its opponent's successes. It now replaces its earlier valuations by the new ones and goes on playing as a new and better machine. *Such a machine would no longer have a rigid personality, and the tricks which were once successful against it will ultimately fail* (*Cybernetics* 172; emphasis mine).

Now, both the human and the machine are capable of error, but it is precisely this contingency which allows both to organize themselves as increasingly strong players, that is, to learn. Like the human, then, the chess-playing automaton "will continually transform itself into a different machine" (*God and Gollum* 21) through the recursive interplay between deterministic rules and a contingent rewriting of those rules.

In this scenario, where both the human and the machine "may absorb in the course of time something of the policy of its opponents" (*Cybernetics* 172),[188] these separate self-organizing systems simultaneously participate in a higher-order chess-playing system where both the ontological and conceptual oppositions between the human and machine dissolve into the pure function of contingency.

Turing's (Thinking-)Machines

For Turing, perhaps even more than for Wiener, questions aimed at identifying ontological or conceptual boundaries between the human and the machine were "too meaningless to deserve discussion" (Turing "Computing Machinery and Intelligence" 449). Turing, nevertheless, did address such questions often and at length. However, as with Wiener, Turing's material and discursive multiplications of the thinking-machine effected a transformation of its grounding concepts. Thus, the question which occupied Turing most urgently throughout his career, "Can machines think?" became instead a multiplicity of iterations describing the more or less contingent quality of both machine and human function--such as, for instance, one's ability to successfully play the Imitation Game (Turing "Computing Machinery and Intelligence" 441).

As is generally known, Turing's second published paper in mathematics, appearing late in 1936, described an "abstract digital computing machine," or "universal computing machine" (Copeland 1; Turing "On Computable Numbers" 68), and the class of problems which it could be designed to solve--specifically, "those calculable by finite means" (Turing "On Calculable Numbers" 58).[189] Turing's main innovation in this paper was the radicality of his emphasis upon computer programming over hardware, as the "machine" thus described was nothing more than a system of rules for manipulating symbols in order to obtain determinate (but not always predictable) results.[190] Imagined as an infinitely long strip of paper, the real nature of the machine was understood as a function of its organization, or programming.[191]

Thus, importantly, what is "universal" about the universal Turing machine is that it can *become* any other Turing machine by virtue of having a description of that machine programmed onto its tape. That is, as Turing would explain over a decade later, "[w]hen we have decided what machine we wish to imitate we punch a description of it on the tape of the universal machine" ("Lecture on the Automatic Computing Engine" 383). In this way, the universal Turing machine is already, in this early paper, participating in a kind of Imitation Game like the one that will later serve as the basis for the famous

Turing Test for machine intelligence. As with that later test, for Turing, ontological questions have already been transformed into questions about function in this 1936 essay.

Turing describes his abstract machines as "automatic machines," of which it can be said that "at each stage the motion [. . .] is *completely* determined by the configuration" and requires no intervention on the part of an "external operator" ("Computable Numbers" 60).[192] This automatic machine consists of a tape divided into squares, each of which bears a symbol (1 or 0).[193] For any state, one square is scanned, and this can result in the machine performing one of a number of possible actions, for example erasing the square, writing a new symbol on the square, or moving the tape left or right in order to scan a new square (Turing "On Computable Numbers" 59).

Thus, any computable number--and, by extension, any operation which can be described in terms of computable numbers--can be represented as a universal Turing machine. For Turing, this is the same as its *becoming* a universal Turing machine. Furthermore, by combining these basic operations, an abstract universal Turing machine can achieve *any level of complexity desired*, given enough time and memory to process suitable instructions (Turing "On Computable Numbers" *passim*).

Two years after publishing this ground-breaking paper, Turing earned his doctorate in mathematics at Princeton University with a dissertation that continued his study of the "uncomputable" problems identified in "On Calculable Numbers."[194] These "unsolvable" problems ultimately lead the machine into a closed loop which makes it impossible for the machine to give a correct answer. Turing would later suggest that rather than allowing the machine to fail in giving an answer, it should be programmed to give a randomly incorrect answer. This ability to err (and to learn from error) is, Turing believed, one necessary condition of thought, or intelligence. Thus, Turing's earliest work on computers was not only based in the formal reasoning language of mathematics, but also struggled with the relationships between deterministic and contingent function.

In 1939, Turing went to work designing and programming machines for the purpose of deciphering the ENIGMA[195] code at Bletchley Park, about fifty miles north of London (Copeland 1-3; Bell 13-15), where he worked and socialized with "the best mathematicians, linguists, and electronic engineers [. . .] plus the two best chess players in the country" (Bell 14). Turing's contribution to the war effort was only declassified in 2004, and turns out to have been much more central than generally suspected. He is now credited with breaking the Naval Enigma, which had decisive consequences for the Battle of the Atlantic, and as the principle architect of the "bombe" code-breaking machine, "which produced a flood of high-grade intelligence from Enigma." These contributions, according to the assessment of British Secret Service official historian, Sir Harry Hinsley, shortened the war by about two years (Copeland 2; 218).

In addition to designing the "bombe" computers, which eventually broke the Naval Enigma, Turing also developed a hand-calculation system called "Banburismus" to assist the bombe machines, broke the Tunny code by a process dubbed "Turingery" by his colleagues, spent time in the United States consulting with American code-breakers, and finally returned to Bletchley Park to work on an automatic system for enciphering speech (Copeland 262-263).

It is also worth noting that, while at Bletchley Park, Turing circulated a typescript addressing machine intelligence, probably focusing upon problem-solving through mechanical search methods, and machine learning through feedback and experience. Both of these functions were important aspects of the way his bombe machines worked as code-breakers (Copeland 351). In Turing's early work, then, he was already treating problems of mathematics, of intelligence or thought, and of language as multiple iterations of a new problematic emerging around the related functions of determinate rules and their contingent unfolding.

Moreover, although Turing himself was reputedly "a complete duffer at the game" (Bell 15), his interests at Bletchley Park already seem to have connected chess with language, logic, thought, and computing as iterations describing this same problematic--and to see that problematic as a transformation in which ontological questions about human or machine identities fall away as unintelligible.[196] Indeed, it was while at Bletchley that Turing first began work on a chess-playing machine, and Bell suggests that "there seems little doubt that the [contemporary] man-versus-machine chess-playing argument really began at Bletchley with Turing playing a major role" (Bell 15).

Donald Michie, a colleague at Bletchley Park reports that Turing also discussed applying the same methods of search and machine learning through feedback and experience developed for the bombe to the problem of chess-playing machines (Copeland 353). Another colleague, I. J. Good, recalls that some of his discussions with Turing at Bletchley Park

> were concerned with the possibilities of machine intelligence, and especially with automatic chess-playing. We agreed that the most interesting aspect of this topic would be the extent to which the machine might be able to simulate human thought processes (qtd. in Bell 14).[197]

Good, in fact, credits himself and Turing with "the notion of a true search, with truncation and evaluation at quiescent positions [. . .] long before Shannon's paper was published" (qtd. in Bell 14).[198] However, Turing was a nexus as much as he was a "great man," and, R. V. Jones, whose work against the German radar defense system was closely connected to Turing's work on ENIGMA, suggests that "[a]s regards the connection between chess playing and decipherment, this was very conspicuous to some of us during the war" (qtd. in Bell 14).

Turing worked on the chess problem steadily throughout his career, devising a number of paper machines based on "rule of thumb" play, and somewhat obsessively playing games against them. By 1951 or so, he had developed TuroChamp with an old Bletchley Park colleague, David Champernowne, and had played at least one, but probably several, partial games against other paper machines. Sometime in 1951 or 1952, with little fanfare, Turochamp and Alick Glennie, a "weak player who did not know the system" (Turing; "Chess" 573; Bell 17),[199] played the first full formal game of chess between a human being and a machine.[200] Glennie, playing black, won in twenty-nine moves (Bell 14-23). Though arguably of less historical importance, it is clear that Turing's interest in chess-playing machines was, like mathematics and code-breaking, a variety of his interest in the limits and abilities of Turing machines, and their simulation of human thought processes.

In 1945, John Womersley recruited Turing to join the Mathematics Division of the National Physical Laboratory (NPL) in London (Copeland 363). Credited as "the first relatively complete specification of an electronic stored-programme digital computer," Turing presented his "Proposal for the Development of an Automatic Computing Engine (ACE)"[201] to the NPL in February 1946 (Copeland 363-365). As in his 1936 paper, "On Computable Numbers," the most significant feature of Turing's proposal, in contrast to the small handful of other contemporary designs for electronic digital computing machines, was its emphasis on software rather than hardware.[202]

Despite its status as the first detailed plan for an electronic digital computer, however, progress on the actual building of ACE was slow and tentative, and Turing left the project in 1948 (Copeland 367) to join M. H. A. Newman at the University of Manchester, where a computer based on Turing's universal computing machine was already well under way (Copeland 367; 373; 395-401).[203]

While on leave from ACE, and shortly thereafter at Manchester, Turing began theoretical work on the problem of mechanizing "higher" brain functions, including the ability to "learn by experience" (qtd. in Copeland 400).[204] The resulting report, rejected by his superiors, and unpublished during his lifetime, outlines a plan for artificial neural nets, anticipating (or inventing) the concept of connectionism that dominates contemporary AI and complexity theory (Copeland 401).[205] Once at Manchester, with a working stored-program computer finally at his disposal,[206] Turing turned his attention to the related problem of Artificial Life (AL) (Copeland 401),[207] which occupied him until his death in 1954.

Turing's Tests
In his 1947 "Lecture on the Automatic Computing Engine," Turing indirectly quotes what is probably Ada Lovelace's most pessimistic comment on the subject of thinking-machines from her notes on Menabrea. "It has been said," Turing writes, "that computing machines can only carry out the processes that they are instructed to do" (392-393).[208] He then proceeds to counter this argument with a passage that could almost have been taken directly from Babbage:

> Let us suppose that we have set up a machine with certain initial instruction tables, so constructed that these tables might on occasion, if good reason arose, modify those tables. On can imagine that after the machine had been operating for some time [. . .] it might still be getting results of the type desired when the machine was first set up, but in a much more efficient manner. In such a case one would have to admit that the progress of the machine had not been foreseen when its original instructions were put in. [. . .] When this happens I feel that one is obliged to regard the machine as showing intelligence ("Lecture on the Automatic Computing Engine" 393).

In conversations, too, Turing described his work on ACE as "building a brain" (Copeland 374), and this imagery forms the basis for his letter to cyberneticist W. Ross Ashby in 1946, in which he admits what must have been clear to anyone working with him on the practical computing problems: "I am more interested in the possibility of producing models of the action [function] of the brain than in the practical applications of computing" (Turing "Letter to W. R. Ashby"). Turing goes on to write:

> The ACE will be used [. . .] in the first instance in an entirely disciplined manner, similar to the action of the lower centres [. . .] it will be entirely uncritical [. . .] devoid of originality. There is, however, no reason why the machine should always be used in such a manner [. . .] It would be quite possible for the machine to try out variations of behaviour and accept or reject them in the manner you describe [. . .] Thus although the brain may in fact operate by changing its neuron circuits [. . .] we could, nevertheless, make a model, within the ACE, in which this possibility was allowed for ("Letter to W. R. Ashby").

Again, in the same letter, Turing implies his understanding of the human brain as a particularly complex example of the universal machines described earlier in his career. "The ACE," he writes, "is in fact, analogous to the 'universal machine' described in my paper on computable numbers."[209]

Turing repeats this line of thought a few years later. Having established that a "digital computer is a *universal* machine in the sense that it can be made to replace any machine of a certain very wide class" ("Can Digital Computers Think?" 482-483), Turing goes on to argue that

> [i]f now some particular machine can be described as a brain we have only to programme our digital computer to imitate it and *it will also be a brain*. If it is accepted that real brains, as found in animals, and in particular men, are a sort of machine, it will follow that our digital computer, suitably programmed, *will behave like a brain*" (483; emphasis mine).

It is significant that Turing uses "will be" and "will behave like" precisely synonymously in this passage, since for him ontology is nothing other than a description of behavior, or function. And, in this way, Turing's tautological--or *a priori*--proof in defense of thinking-machines--based in a concept of thought that is entirely and admittedly functional rather than ontological--neatly reverses the tautological and "*a priori*" arguments of Descartes and Poe. Thus, while within a Cartesian ontology a thinking-machine is inconceivable because, by definition, a machine is not a thinking-thing, for Turing, if it is doing what we normally call thinking, and if it is what we normally call a machine, then we might as well call it a thinking-machine.

Importantly, for Turing it is also this functional concept of thought (as contingency) that provides the basis for the Turing Test for machine intelligence. In short, the Turing Test posits that we should ascribe to a machine which appears to function intelligently, actual intelligence. This argument is also a transformation of the problem commonly referred to by analytical philosophers as the "problem of other minds." Turing's suggestion is that we ought to treat the problem of computer minds exactly as we routinely treat the problem of other (human) minds--that is, as a practical expediency, in the absence of reliable means for proving a negative, we ought to accept them at face value.

In other words that closely paraphrase Descartes, "the only way by which one could be sure that a machine thinks is to *be* the machine and to feel oneself thinking [. . .] Likewise [. . .] the only way to know that a *man* thinks is to be that particular man" (Turing "Computing Machinery and Intelligence" 453). That is, only the machine itself can think and therefore know that it is a thinking-thing. But, while we can never know for certain that *any* other mind exists, "it is usual to have the polite convention that everyone thinks" (Turing "Computing Machinery and Intelligence" 441), and there is no reason for Turing why this policy should not apply to any apparently-thinking-thing.

Similarly, in his last published essay, in response to objections, Turing addresses the question, "Could one make a machine which would have feelings like you or I do?" ("Chess" 569). Turing's playful answer is precisely the same as his answer to the question of machine intelligence, and in effect, a repudiation of the meaningfulness of the question: "'I shall never know, any more than I shall ever be quit certain that *you* feel as I do'" (569).[210] Thus, for Turing, the test for thinking-machines is precisely the test for any other thinking-thing, while, as always, questions about ontology and substance are dismissed as confusing or meaningless.

Of course, if machines (like humans) can be said to think simply by virtue of their appearing to think, then some test is required for the appearance of thought. Importantly, for Turing, these tests of intelligent function always involved some variation of contingently applied (deterministic) rules of play-- such as in conversation or chess.[211]

Indeed, it seems almost inevitable that Turing's best-known discussion of the Turing Test for thinking-machines involved not only mathematics, but both speech *and* chess as well.[212] Thus, in 1950, Turing offers this sample conversation during a hypothetical Turing Test:

Q: Please write me a sonnet on the subject of the Fourth Bridge.
A: Count me out on this one. I never could write poetry.
Q: Add 34957 and 70764.
A: (Pause about 30 seconds and then give as answer) 105621.
Q: Do you play chess?
A: Yes.
Q: I have K at my K1 and no other pieces. You have only K at K6 and R at R1. It is your move. What do you play?
A: (After a pause of 15 seconds) R-R8 mate. ("Computing Machinery and Intelligence" 442).

As early as 1948, however, Turing discussed in an internal report for the Manchester group, a chess-playing version of the Imitation Game:

It is not difficult to devise a paper machine which will

play a not very bad game of chess.[213] [. . .] A and C are to be rather poor chess players, B is the operator who works the paper machine. [. . .] Two rooms are used with some arrangement for communicating moves, and a game is played between C and either A or the paper machine. C may find it quite difficult to tell which he is playing" ("Intelligent Machinery" 431).

Turing goes on to note that this "is a rather idealized form of an experiment that I have actually done" (431).[214]

The passage resonates strongly with the fuller treatment that Turing gives the Imitation Game in "Computing Machinery," and strongly suggests that his inclusion of a chess problem there was anything but accidental. It is not only the inclusion of the chess problem in the latter essay that is striking, but the way in which it is imbedded within "ordinary" speech, or conversation, as yet another iteration of intelligence as the self-organizing system between two actors within a mutually productive field of determined rules and contingent play.

Indeed, as we have seen, for Turing, the problems of machine intelligence (AI), machine language, and chess-playing machines were all problems of programming universal Turing machines. That is, a chess-playing machine, a machine capable of human-like speech, or any other thinking-machine would necessarily be a description of a Turing machine. Furthermore, as we have also seen, a human being capable of all these functions would be one description of a universal Turing machine. Considered this way, the Turing Test for machine intelligence becomes a means for humans to test *other* Turing machines.[215]

Thus, the ACE, for instance, was theoretically capable of reproducing the functions of the human brain precisely because, as his paper "On Computable Numbers" argued, any universal Turing machine of sufficient capacity can be made to imitate any other universal machine. And, for Turing--in that paper and elsewhere--to imitate is to *become*. Thus, already implicit in Turing's thinking in 1946, and elaborated repeatedly throughout his career, is the idea of a relationship between intelligence and contingency in the form of self-organizing systems of learning, and the conception of a human brain as just such a self-organizing universal machine.

Turing's Transformation

For Turing, as for La Mettrie, everything was mechanism; but this did not mean that for Turing everything was "automatic" or determined. Rather, Turing suggested, some of the most interesting machines have the ability to function contingently, to learn from experience, and thus to rewrite their deterministic programming--that is, they are intelligent and self-organizing. A human being, Turing believed, was one example of such a self-organizing machine--but not the only kind conceivable.[216] That this kind of behavior is at odds with our inherited concept of both the human and the machine did not escape Turing, and despite his frequent use of these concepts in his writing, he just as often challenged their usefulness or meaningfulness for the questions that they attempt to pose.

Thus, while Turing continued to use the language of men and machines, he transformed them in provocative ways that called into question their status as meaningful concepts or substantially separate entities, often quite directly. For instance, the first sentence in his 1950 essay in *Mind* states, "I propose to consider the question, 'Can machines think.'" The second sentence objects that this consideration "should begin with definitions of the meaning of 'think' and 'computer.'" But, Turing goes on to say, if "the meaning of the words 'think' and 'machine' are to be found by examining how they are commonly used, it is difficult to escape the conclusion that the meaning and the answer to the question, 'Can machines think' is to be sought in a statistical survey such as a Gallop poll" ("Computing Machinery and Intelligence" 441).

In other words, as we have seen, there is an *a priori* argument against the conceivability of a thinking-machine, given the dualist Cartesianism that grounds the concepts that formulate the question, "Can a Machine think?"[217] Rejecting the project of redefining 'machine' and 'think' to suit the reality of functioning machines,[218] Turing decides instead to "replace the question by another" ("Computing Machinery and Intelligence" 441). That question, of course, takes the form of the Imitation Game.

Indeed, throughout his writing, Turing seems to be as concerned with the limits of the language and concepts through which such questions as "Can machines think?" are posed, as he is with the answers to those questions. And, much like Wiener, he concludes that the questions themselves, to the extent that they are grounded in these concepts, are "too meaningless to deserve discussion" (Turing "Computing Machinery and Intelligence" 449).[219]

Specifically, Turing challenges the notion of a real ontological or meaningful conceptual distinction between the human and the machine, as well as, ultimately, the Cartesian concept of thought that grounds it. In other words, Turing transforms the discourse surrounding machine intelligence, and the image of the thinking-machine, by stressing the functional play between determinate rules and contingent behavior that organizes systems toward specific goals through "mental" processes like search, testing, evaluation, and learning (memory).

For example, in his 1947 lecture on ACE to the London Mathematical Society, Turing directly confronts what he takes to be misleading or nonsensical language. Addressing the ontological paradox inherent in the image of a thinking-machine by connecting intelligence with the play between determinacy and contingency, he defines learning as a function of self-organization, and, in the process, redefines the problematic in functional terms. "It might be argued," he writes,

> that there is a fundamental contradiction in the idea of a machine with intelligence. It is certainly true that 'acting like a machine' has become synonymous with lack of adaptability. But the reason for this is obvious. Machines

in the past have had very little storage, and there has been no question of the machine having any discretion. [. . .] In other words, then if a machine is expected to be infallible, it cannot also be intelligent" ("Lecture on the Automatic Computing Engine" 393-394).

Notably, Turing begins the passage with an acknowledgement of the inconceivability of a thinking-machine, given the Cartesian implications of its component terms, "thinking" and "machine"; but, typically, Turing refigures this essential ontological opposition between human and machine as a practical functional difference between intelligent and non-intelligent machines. And this opposition is just the difference between a machine that is completely determined by its initial programming, and one that is able to adjust, or to (re)organize, its own programming in response to experience (that is, feedback). Thus, when Turing claims that "the ACE can be made to do any job that could be done by a human computer" ("Lecture on the Automatic Computing Engine" 378), he means it in the most radical sense possible.

In his 1948 essay "Intelligent Machinery," Turing makes the significance of this interplay between determinism and contingency explicit, describing intelligence as emerging from the self-organizing recursive relationship between the determined rules of a machine's programming and "interference" from its environment ("Intelligent Machinery" 421-422). Describing "the cortex of the infant [a]s an unorganised machine, which can be organised by suitable interfering training" ("Intelligent Machinery" 424), Turing goes on to detail the mechanism through which unorganized machines might organize themselves.

In short, the process has to do with programming the computer to generate occasional random answers under certain conditions (for example, when a determined answer is not possible, as in Gödel's theorem, or in the absence of specific programming), which it will then fold into its existing program. That is, when "a configuration is reached for which the action is undetermined, a random choice for the missing data is made in the description, tentatively, and is applied" ("Intelligent Machinery" 425). In other words, *the machine's capacity to generate undetermined behavior is determined by its programming.*

The machine "learns" by evaluating the effectiveness of the random choice toward achieving some goal, which may be judged in relation to "pleasure' or "pain" (Turing "Intelligent Machinery" 425). In this sense, "intellectual activity consists mainly in various kinds of search," and it is the machine's relative ability to organize itself toward some goal that constitutes intelligence--whether that goal is to solve a mathematical problem, decode encrypted text, acquire linguistic or social skills, check-mate an opponent, or achieve genetic survival. In other words, thought, for Turing, is certainly not a substance. To the extent that it holds up as a concept at all, it describes the play between determined rules and contingent function, the emergent cycle between order and chaos.

Thus, Turing writes, "My contention is that machines can be constructed which will simulate the behaviour of the human mind very closely. They will make mistakes at times, and at times they may make new and very interesting statements" ("Intelligent Machinery, A Heretical Theory" 473). In this essay, Turing directly addresses the presumption that intelligence is a proprietary domain of the human based on the human capacity for contingent behavior, and the machine's deterministic, "automatic" programming:

> By Gödel's famous theorem, or some similar argument, one can show that however the machine is constructed there are bound to be cases where the machine fails to give an answer, but a mathematician would be able to. On the other hand [. . .] the mathematician makes a certain proportion of mistakes. I believe that this danger of the mathematician making mistakes is an unavoidable corollary of his power of sometimes hitting upon a new method" ("Intelligent Machinery, A Heretical Theory"

472).

Again, Turing's solution to the apparently insurmountable limit of deterministic programming is to incorporate an element of randomness into that programming.[220] That is, in cases where a universal Turing machine encounters a mathematically unsolvable problem, and thus should give no result at all, Turing proposes that the machine be programmed to give a randomly incorrect answer instead--that is, to err. In this sense, a thinking-machine is no longer recognizable as a "machine" in Cartesian terms--that is, it is not "'acting like a machine'" (Turing "Lecture on the Automatic Computing Engine" 393-394).

Turing states the point this way in "Computing Machinery":

Most programmes which we put into the machine will result in its doing something that we cannot make sense of at all, or which we regard as completely random. Intelligent behaviour presumably consists in a departure from the completely disciplined behaviour involved in computation, but a rather slight one, which does not give rise to random behaviour or to repetitive loops (463).

Thus, again, while Turing plays with ontological language and concepts, he is not, in fact, concerned with what humans or machines *are*; he is, instead, strictly interested in what they do and how they work. Like Descartes, he locates human intelligence within a capacity for contingent function, and like Poe, he casts this contingency as error. But, for Turing, this situation is itself nothing more significant than a contingency of history--a result of the simple fact that non-human machines with such functional capabilities did not as yet exist. Thus, for Turing, the exercise of deciding between machine or human identity is ultimately pointless. What interests Turing is the functional processes by which a deterministic set of rules can (re)organize itself to respond in new and unexpected ways, that is, contingently.

The same kind of transformation occurs in relation to a variant of the "Can machines think?" question that Turing spent a lifetime (not) answering, that is, "Can the machine play chess?" (Turing "Proposed Electronic Calculator" 16). By the time Turing addressed this question within the context of his work on electronic calculating machines, he had already been working on the same problem with pencil and paper for several years (Bell 17; Copeland 353). Thus, Turing's earliest efforts, like those of his Bletchley colleagues, involved only the program for deciding moves within the legal rules of play--that is, for ordering contingent function. Though he did not name either, his earliest extant written answer[221] to this chess problem seems to directly address the technological question raised by Kempelen in the specific ontological terms answered by Poe.[222] That is, Turing understood the problem in terms of broad philosophical questions as much he did in terms of narrow technical ones, and specifically as a problem regarding the relationship between contingency and determinism as they relate to the function of reason (computation), and to the ontological limits of the machine.

Indeed, for Turing, the philosophical and technical problems are not really separate questions, but rather each is a partial formulation of the functional problematic. Thus, as he will repeatedly do in future discussions of chess-playing machines and machine intelligence generally, Turing transforms both the question and the answer by abandoning their ontological grounds. "A machine," Turing writes,

> could fairly easily be made to play a rather bad game [of chess]. It would be bad because chess requires intelligence. We stated at the beginning of this section that the machine should be treated as entirely without intelligence.[223] There are indications however that it is possible to make the machine display intelligence at the risk of its making occasional serious mistakes. By following up this aspect the machine could probably be made to play very good

chess" ("Proposed Electronic Calculator" 16).
In other words, again, it is precisely contingency, the possibility of error--that is, the possibility of contingently deviating from determined programming--that defines intelligence, as expressed by chess-playing, for Turing. In this sense, the thinking-machine is precisely *not* an automaton as understood by Descartes or Poe.[224]

But, Turing's transformation of questions about the ontological and conceptual limits of humans and machines also raises questions about the concepts of intelligence and thought themselves. The point is underscored in a provocative section of "Intelligent Machinery," where Turing seems to suggest that intelligence is more a relation than a property or substance. "If we are unable to explain and predict [an object's] behaviour," he writes,

> or if there seems to be little underlying plan, we have little temptation to imagine intelligence. With the same object therefore it is possible that one man would consider it as intelligent and another would not; the second man would have found out the rules of its behaviour" (431).

In other words, Turing explicitly recognizes that both an initial set of deterministic rules *and* an element of randomness or contingent response to "interference" is necessary for the development of any self-organizing system, such that *either* too much chaos *or* too much order will limit the emergence of intelligence. But, he also recognizes that such systems are always imbedded within other dynamic systems. Thus, not only is intelligence figured as a functional balance between contingency and deterministic rules, but it is also relational, or relative, to other dynamic systems.

Conclusion

For both Wiener and Turing, then, Descartes' claim that a thinking-machine is inconceivable points to the limits of our concepts rather than to the limits of machines. Both conceived the thinking-machine through multiple iterations not only as a class of material mechanism, but also as the multiple images building up the details toward a new conceptual schema which replaced ontology with relation and function. For Wiener and Turing, the trope of the thinking-machine in its multiplicity thus effected a detachment of the component concepts identifying and opposing the human and the machine within a Cartesian ontology, reshuffling and redistributing those components across boundaries in unexpected configurations. Thus, not only could Wiener and Turing write of chess-playing-machines and mechanical brains, but also of human and social machines, DNA-machines, human-airplane machines, machines-that-do-not-act-like-machines, and infant neural-nets-as-machines.

In other words, the images of the thinking-machine, multiply iterated as the chess-playing machine, the speaking-machine, the reasoning-machine, and (as all of these) the human-machine--within and against a Cartesian metaphysics, where determinism and contingency are understood as being essentially identified with ontologically opposed substances--began to resolve into concepts of a new problematic, which no longer recognized the incompatibility of mind and body, but sought rather to understand how (within one substance) complex contingent behavior could emerge as the function of simple deterministic rules. For both Wiener and Turing, mathematics, reasoning, language (as both cipher and speech), and chess all expressed ways into this same problematic.

For both Wiener and Turing, this new problematic also demanded a new language and new concepts. Thus, struck by the inadequacy of the inherited categorical oppositions between human and machine, vitalism and mechanism, determinism and contingency, and the language that invokes them, both Wiener and Turing explicitly sought to reject these inherited concepts and questions in favor of a new language and new concepts with which to describe the new problematic.

Thus questions about the function of the thinking-machine emerged as the discursive transformations of questions about its substance and being. Furthermore, while the play between determinism and contingency remained a core feature of this discourse, and while contingency remained the privileged term, the emerging concepts of cybernetics and self-organizing systems recognized a mutually productive relationship between determinism and contingency, as well. In other words, contingency was no longer merely opposed to determinism as a uniquely human substance, but, rather, was understood as an emergent function of any set of deterministic rules of the right kind. The concept of the human drops out of the discourse because its defining content--the function of contingency--has been redistributed across ontological boundaries.

Thus, paradoxically, for Wiener and Turing, in the moment that it becomes possible to conceive of, and to meaningfully say, "thinking-machine"--to answer "yes," to questions such as, "Can machines think?"--the questions and the concepts which ground them have already become meaningless. The conceptual components of the human have been folded into the concept of the machine, and the ontological content of the human category has become conspicuously empty. As thinking-machines, then, we are only contingently human, which is to say, posthuman; and despite himself, it is through these transformations of Descartes that the posthuman is conceived. In this sense, we are all Descartes' daughters.

CONCLUSION: IDENTITY AND COMPLEXITY

> "[I]dentity is not what matters" (Derek Parfit 216-217, 282).

N. Katherine Hayles opens the first chapter of *How We Became Posthuman* with a narrative describing her own uncanny encounter with the image of the thinking-machine. While engrossed in an otherwise enjoyable account of his robots, Hayles reports, she is "shocked into awareness," recoiling in horror upon reading Hans Moravec's speculations on the possibility of downloading human consciousness into computers. "How," Hayles asks, "was it possible for someone of Moravec's obvious intelligence to believe that mind could be separated from body? [. . .] that consciousness in an entirely different medium would remain unchanged?" (1).

How, we might ask, is it possible for someone of Hayles' obvious expertise on the subject to so misread Moravec by attributing to him these views? And, how could someone of Hayles' stature use such a misreading to characterize the project of AI and the emergence of the posthuman generally? Indeed, Hayles' reading of Moravec is especially perplexing, since his signature engineering advances in robotics have focused on the development of sensory inputs, and his writing has theorized and defended a dynamic developmental relationship between the contingencies of embodiment and the actualization of higher thought processes.[225] Moreover, as we have seen, the understanding of body and mind as interrelated and mutually productive organizational processes within larger complex systems has been a key feature of AI engineering and theory from its beginnings,[226] and, as I am arguing, has powerfully transformed the Cartesian dualism upon which human identity, and identities generally, are grounded.

Moreover, Hayles' critique of cybernetics and systems theory fails to acknowledge that, for them, at a different level of organization, the "medium" is itself information. That is, this opposition between medium and message, too, is contingent and functional, not deterministic or ontological, and human flesh and silicon chips are materially different *because* they express different sets of information. Thus, cybernetics and AI, as iterations of complex systems theory, conceive information not as a materially independent substance (as Hayles charges), but rather as descriptions of material organizations. That is, information can move, but not without reorganizing matter in the process; indeed, information just is that (re)organization. For this reason (but not for the vitalist reasons that she seems to invoke), Hayles is correct in asserting that a human consciousness downloaded into a computer would not remain "unchanged" (*How* 1).

But, again, that is not quite Moravec's position. Indeed, one may justifiably critique a certain giddy naiveté in Moravec's optimism, but if there is no expected change in the downloaded subject, there is for him no utopian project, and therefore no point to the exercise.[227] What Hayles misses is that it is the transformation and dissemination of the identifying component concepts of the human (namely contingency expressed as thought, speech, etc.) into nonhuman actualizations that allows Moravec to stop worrying about whether or not the human survives, and instead to concentrate on liberating that which is important and positive in our concept of the human from that which is not.

But, even if Hayles has read Moravec's utopian "fantasy" more or less rightly, what could be the nature of a threat so alarming that its appearance could turn Hayles' enjoyable read instantly into a dystopian "nightmare"? (Hayles *How* 1). Certainly, Hayles' admittedly strong emotional reaction to Moravec's thinking-machines suggests that there is more at stake for her than a simple academic disagreement. What, then, is the real motive of Hayles' anxiety, and what is the source of her confusion?

To begin, the logic of Hayles' argument seems to suggest ambivalence. On the one hand, she argues that there can be no human without a human body. In other words, Hayles wants to propose a radical materialism in which to change the "medium" is to change the "message" (1).[228] As noted above, this is not yet inconsistent with either cybernetics or contemporary systems theory. But, there is something strange about Hayles' objection to this change. Indeed, her defense of the human, and of the body in particular, seems to derive from a suspiciously Cartesianist philosophy of *being*. That is, Hayles figures the human body as the singular and *unchanging* substance in which human consciousness alone can abide. But, if her objection to Moravec is that he opens the human to transformation, then her concept of the human, and thus of human subjectivity, must be of a static, closed system unopen to change. Moreover, there is something oddly *inhuman* about Hayles' image of the human here, for it seems to forget the role of contingency as the identifying function of the human, even though that is clearly part of what is important to her about the human. In this way, the dynamic between Moravec and Hayles effects a transformation of Cartesianism, in which the human comes to represent the determined, while the machine signifies contingency. But, Hayles does not seem to be aware of this reversal.

Thus, it appears that Hayles objects not to Moravec's technological achievements (which she finds "enjoyable"), but to their ontological implications. For Hayles, it seems, the "nightmare" of the "roboticist's dream" is precisely the loss of human identity, conceived as a closed system, which, in a strange and twisting line both through and away from Descartes, returns to its essential location within the human body (1). An iteration of a now-familiar pattern, Hayles' anxiety over the loss of ontological categories and the redistribution of their component concepts is expressed narratively as the physical dominance and subjugation of human bodies by machines.

Despite her anxiety, however, Hayles finds the posthuman "fascinating [as well as] troubling" (35). And, it is, after all, the posthuman potential for changing human subjectivities away from the repressive aspects of the so-called liberal subject, that excites and titillates her. For Hayles, the posthuman "evokes the exhilarating prospect of getting out of some of the old boxes and opening up new ways of thinking about what being human means" (285). Here, Hayles' concept of the posthuman seems to express an ontology of *becoming*, where the human system is in a constant state of self-creation in relation to other systems, both human and nonhuman--where, in other words, it is the process of actualization, of becoming, through the interplay of rule and contingency, that changes us, whatever our material embodiments. But, if this is the case that Hayles wishes to make, then the human (body) cannot also be the unchanging medium that she insists upon defending against the violations of Moravec's machines. In fact, it must be a complex system.

Hayles' confusion, like Poe's, seems to stem from her anxious reluctance to let go of the organizing trope of human identity, precisely because she is inclined to infuse it with an ontological status. And, it is just this mistake that gets Hayles into trouble, because it forces her to argue that the human is both unchanging and in flux.

Hayles also effects a similar misreading of Wiener, such that we can begin to detect a pattern to her anxieties. As with Moravec, Hayles suggests that "the values of liberal humanism

[. . .] deeply inform Wiener's work" (*How* 86-108). Indeed, much of Wiener's popular writing does seem to dwell on the now-familiar anxiety about the machine's replacement or mastery of the human. Thus, even before he had coined the term "cybernetics," Wiener reflected that he "had become engaged in the study of a mechanico-electrical system which was designed to usurp a specifically human function" (*Cybernetics* 6). And, it would be easy to think that Wiener had fallen back into a very Cartesianist mindset when he observed that

> the first industrial revolution, the revolution of the 'dark satanic mills,' was the devaluation of the human arm by the competition of machinery [while] the modern industrial revolution is similarly bound to devalue the human brain. [. . .] The answer, of course, is to have a society based on human values other than buying or selling (*Cybernetics* 28).

But, Hayles' reading largely misses the point, for there has been a transformation here, too. Certainly, Wiener does harbor deep anxieties about the potential loss of human access to the privileged functions of contingency, or, to the adaptive, flexible, learning function that makes us "alive"; and in his warnings to this effect, he does often draw upon the Cartesian imagery that essentially links the human with contingency, and the machine with determinism. But, this is not precisely the same thing as defending "the liberal humanist subject," nor does it constitute a "horrified withdrawal" from the libratory impulses of cybernetics or the posthuman (Hayles *How* 86). Rather, it is the effect of the limits of the conceptual and discursive apparatus to which he had access--limits with which he often expressed his profound frustration.

For Wiener, then, "evil" is not so simple as the reversal of positions between human master and mechanical slave, Bierce's Oedipal nightmare of the creature usurping the place of its human creator; and, it is certainly not the loss of the human as an organizing concept. But, it is a transformation of these anxieties, through a redistribution of their core components across the boundaries and *away* from the ontological concepts that ground liberal humanism. Thus, what worries Wiener is the coerced constraint of open, complex systems within the confines of deterministic rules or order--and his examples of such systems are not confined to machines.[229] Wiener uses the human as a trope for contingency, because it is convenient to use the language that is available to him--but he certainly does not give this convenience any ontological force. Indeed, it is precisely because it has no ontological value that the human can stand as a metaphor for the non-rigid, non-deterministic systems that Wiener wishes to defend.

For Wiener, as we have seen, it is the play between order and chaos, iterated as a non-linear algorithm, or transformation, that describes the becoming of any complex system, such as a human or a thinking-machine, over time. Thus, while the ability to respond contingently--that is, to think (or learn)--is no longer for Wiener the exclusive domain of the human--and while all thinking systems require some basic set of rules in order to organize as complex systems--between determinism and contingency, it is contingency, as the motor of becoming, that remains for him the privileged term--the "good" to be nurtured and protected--that about the human which *matters*.[230]

Thus, Hayles' attribution to Wiener of "cybernetic anxiety" is not only a serious misreading of Wiener that fails to comprehend the transformation that his work effects, but it appears to be a textbook example of psychological projection as well. Indeed, as we have seen, Hayles' text is fraught with anxiety concerning the loss of the human, which she oddly equates with a loss of the body within a classic Cartesian dualism. Not surprisingly, Hayles' language is often symptomatic of her inability (or unwillingness) to give up the Cartesian image of thought that she critiques in Moravec and Wiener. And, it is her inability to speak (which is to say, think) the posthuman that disrupts her ability to coherently and fully conceptualize its political potential.

For example, even while she writes hopefully of "the cognisphere" and of "distributed cultural cognitions," in a recent assessment of Haraway's cyborg, Hayles still imagines them as being "embodied both *in* people *and* their technologies" (Hayles "Unfinished" 159-160; emphasis mine). The Cartesianist imagery here is striking on at least two fronts, as she sets up the categories of people and technology as separate and interior spaces to be inhabited by thought, while rejecting discursive moves that would conceive thought as an emergent property actualized or even embodied *as* human and machine. In this way, Hayles speaks *about* systems theory and the posthuman without ever really speaking systems theory or the posthuman; and when she describes the "cognisphere, like the world itself, [as] not binary but multiple, not a split creature but a co-evolving and densely interconnected complex system," we are suspicious of her conviction (Hayles "Unfinished" 165).

For Hayles, then, the posthuman represents the potential liberation of the human from the misogyny, racism, and other violences of a specifically historical liberal subject, but also the potential loss of the human body--and, thereby, of the human. Again, Hayles' confusion seems to stem from her own lingering attachment to dualist ontology, which pervades her reading of the posthuman. Hayles' anxiety about losing the human, then, is grounded in a kind of Cartesianism that takes this dualist ontology for granted, despite her best efforts to think past it. In this strange way, Hayles seems to get repeatedly caught up in the thinking that she attempts to critique. It is Hayles, after all, and not Wiener, whose work is deeply informed by the values of liberal humanism. And, for this reason, it is not precisely the posthuman condition that Hayles dreads, but the *posthumous* condition that attends (for her) the loss of a uniquely embodied human ontology, and therefore of a uniquely human identity. Indeed, Hayles' entire project might be understood as the defensive assertion that "the posthuman does not really mean the end of humanity" (*How* 286).

But, Hayles is not alone in her anxieties. For example,

CONCLUSION: IDENTITY AND COMPLEXITY | 127

in a recent interview by Nicholas Gane, Donna Haraway, whose "Cyborg Manifesto" might be considered *the* germinal text for posthuman political theory,[231] declares herself "deeply resistant to systems theories of all kinds," and characterizes postgender and other post-identity technological, political, and intellectual projects as "blissed out," "stupid and silly" ("When" 139, 146). Insisting upon both the material reality and the socially constructed nature of identity categories, characterizing them as "provisional" and made, but not "made up," Haraway seems to adopt a systems theory approach to ontology, while nevertheless mistrusting attempts to think beyond identities entirely--perhaps confusing them with a belief that understanding a category as constructed will "make it go away" ("When" 153, 137).

Calling instead for "new category work," that is, presumably, for more politically useful and liberating categories still understood as "provisional," Haraway reaches for some way of understanding the "relentlessly relational" constitution of identities, "without throwing the baby out with the bathwater" ("When" 143, 153). Imagining identity categories not "as containers for each other, but as co-constituting verbs," Haraway hopes to escape the tyranny of specific identity categories, but is not yet willing to completely reject their logic ("When" 146).

Thus, while on the one hand recognizing that identities are not "ontologically real and separate" but "constitutive relationalities," and that "it is relationality all the way down," Haraway nonetheless stops short of abandoning the discursive formations that reproduce us as ontologically determined beings. In other words, Haraway wants to keep her interventions on the plane of identity formation, or "worlding," without really intervening in the conceptual systems that render projects of taxonomy sensical ("When" 141-142, 147). Thus, while Haraway's defense of identity categories proves to be much more intellectually subtle and nuanced than Hayles', it nevertheless seems to reproduce the same anxieties, and to effect the same blockages.

But, this attachment to identity bears closer examination, as it has important political and ethical implications. Because it emerges through a transformation of discourses of substance and essence to discourses of function and relation, the deterritorialization of human identity effected through the conception of the thinking-machine and its multiple iterations also implicates the ontological status of identities in general. This can be understood as a general consequence of the posthuman transformation of questions of ontological substance into questions of function and relation, as taken up by complex systems theory. That is, while on the one hand, identity categories like the human and the machine fairly obviously drop out of posthuman discourse as meaningful concepts, social and personal identities, perhaps somewhat less obviously, also tend to lose their ontological status. "We are not stuff that abides," Wiener writes, "but patterns that perpetuate themselves" (*Human Use* 96).

The important force of this line of thought is not to suggest that there is no order or possible ordering of lived experience and felt connections within and between people, but to posit that while these are, indeed, important and indispensable functions that we attach to identity, they are, in fact, functionally independent of identity. The challenge, then, as Haraway suggests, is to find a way of conceptualizing ourselves out of determined identity categories (e.g. human, woman, or person) without losing what matters. But, against the advice of both Haraway and Hayles, it seems that the best way to do this is to abandon our attachments to identities, and to develop instead discursive practices for ordering and analysis which draw upon the conceptual transformations effected through iterations of the thinking-machine, as expressed in complex systems theory. But, this will require a more subtle understanding of what matters about identity than either Hayles or Haraway provide.

This is because, while both Hayles and Haraway locate what matters about identity within concepts and discourses of identity, and therefore conclude that it is identity itself that matters, it is clear that it is *not* in fact identity *per se* that really matters to them. Moreover, it is only within a dualist concept of the human that the thinking-machine marks the end of human identity, and that the end of (human) identity marks the end of what matters. Thus, it is the deep persistence of this dualism--even among those who purport to reject it--that perpetuates anxiety over the end of both the human and identity as meaningful organizational concepts--an anxiety that still blocks contemporary intellectual and political encounters with the posthuman.[232] But, this anxiety, as we have already suspected, belies the persistence of a Cartesianism which grounds all of these oppositions between categories: between human and machine, between body and mind, between master and slave.

   Thus, a rejection of the discourses of identity does not entail the abandonment of projects of ordering, critique, and analysis, but rather a redistribution of what is important and productive about these projects into more coherent and productive systems for doing them. And this is possible precisely because what matters about human identity has been redistributed across emerging concepts of the posthuman and complexity theory, through (among others) the transformations that we have traced as iterations of the thinking-machine.

   The results of these transformations, as we have seen, are two-fold. On the one hand, with the identifying functions and conceptual components of the human appropriated by and redistributed over an increasingly multiple field of material actualization (including both the machine and the animal, which had once marked its boundaries), the concept of the human becomes empty, and its ontological status becomes null, as there is no longer any conceivable single substance upon which to pin it. In this sense, the conception of the thinking-machine indeed marks the death of the human--but, again, *it does not mark the death of what matters.*

That is, in order to understand how "what matters" not only survives, but also multiplies and proliferates, it is necessary to understand the transformation which frees the identifying functions of the human from the human, and actualizes them through iterations of the thinking-machine. Neither the human nor the machine survives this transformation, but the ethical imperative of liberating contingency from closed systems of (human) identity does--or might, if we take enough care. [233]

But, the destructive functions of identities in complex-systems cannot be overstated. No matter how much we multiply or how far we expand the inclusivity of our categories of identity, no matter how much contingency we build into our revisions of identity concepts, as long as we *are* anything, we are not *becoming* anything new, and we remain closed to understanding, or intervening within, the dynamic creative patterns that produce us. All our lines of flight are inevitably recaptured, falling back into the crushing orbit of the "I," or, of the "We."

As we have also seen, it is not enough to simply oppose contingency to order. There is no self-organization, no becoming, without the interplay between contingency and determinism, chaos and order. But, different systems of ordering may give more or less play to contingency, and it is the inventive, productive function of contingency that drives becoming forward. For that reason, our choices about the systems of ordering that we adopt philosophically and discursively have profound political and ethical implications. Thus, I am suggesting that rather than continue to reproduce in our thinking and discourses the concepts of category and identity that function to contain difference and contingency, and with them the potential for remaking our worlds, we should instead work toward new discursive systems of ordering that give priority to multiplicity, relationality, and becoming.

What I am defending is not unlike what Rosi Braidotti calls "process ontology" ("Posthuman" 197). A corollary to her nomadic subject, process ontology can be thought of as an ordering system that "posits the primacy of relations over substance," while stressing the materiality of those relations ("Posthuman" 199). In this ontology, an "entity" designates what would be better characterized as a complex system "stable enough to sustain and to undergo constant [. . .] fluxes of transformation" ("Posthuman" 2001). And, it is not as an entity, but as such a self-organizing complex system, that Braidotti seems to understand subjectivity when she writes that it

> both receives and recomposes itself around the onrush of data and affects. . . . coincid[ing] with nothing more than the degrees, levels, expansion and extension of the head-on rush of the 'outside' inwards. What is mobilized is one's capacity to feel, sense, process and sustain the impact with the complex materiality of the outside (*Transpositions* 145).

Putting this in Deleuzean terms, she continues,
It comes down to a question of style, but style here is no mere rhetorical device, it is rather a set of material coordinates that, assembled and composed in a sustainable and enduring manner, allow for the expression of the affectivity and the forces involved. They thus trigger a process of becoming (*Transpositions* 145-146).

While concepts like *entity* do not prove any more helpful than do categories or identities in liberating difference and contingency from the determinist forces of taxonomy, and although it seems that Braidotti, like Hayles and Haraway, harbors an unhealthy nostalgia for identity categories, nonetheless, Braidotti's use of Deleuze's concept of style to express the productive and transformative patterns of relations between chaos and order seems to go a long way toward throwing out the bathwater without much disturbing the baby.

This is because, just as the posthuman marks the transformation and dissemination of the constitutional components of Cartesian human identity (e.g. contingency) outside the rigid boundaries of the human, so too, complex systems theory allows us to transform and redistribute the component concepts of identity that build common cause and positive political action while leaving the reproductive force of identities and categories behind. Thus, for example, rather than speaking of gender identities, we might instead speak of emerging gender patterns and styles, of self-organizing gender systems; rather than *identifying as* a woman, I might instead *recognize* participation in a pattern of gender iteration that orders other relations with and without my local subjectivity. Similarly, "I" am at any moment not "I," but a specific local nexus of multiple, extended, and interrelated systems of race, class, gender, geography, education, and so forth.

In other words, understanding that subjectivities are constituted as a functional set of actualized material relations does not, in fact, require a concept of the body, or of embodiment; nor does recognizing patterns and styles among subjectivities require taxonomical categories or the concept of identity. Indeed, one way of expressing my argument would be to insist that dynamic patterns should be recognized and critiqued, but not identified. That is, I am suggesting that we reinvent the ways we conceive and speak of these relational materialities, and thus also the ways that we produce them.

The strength of conceptualizing subjectivity(ies) in this way, is that it makes it possible to sidestep the repressive reproductive forces of identities, while retaining both the organizational and analytical connections and continuities demanded by Haraway and the material specificities demanded by Hayles. These, I would suggest--not identities themselves-- are what (should) matter to Hayles and Haraway, and for the productive political interventions that they envision.

	Thinking-Machine TimeLine			
Ramon Llull	1305	*Ars Magna*		
Descartes	1637	Discourse on Method		
Vaucanson	1740 c	Flute-player, Duck		
Wolfgang von Kempelen	1770	Automaton Chess-Player	Queen Maria Theresa Austria-Hungary	
Louis Dutens	1770	Letter to editor describing chess-player	*Le Mercure de France* (sets pattern of debate)	
Kempelen in Paris	1783	Turk plays Philador, Franklin		
Kempelen in London	1873	Philip Thicknese proposes concealed child; Henri Decremps says dwarf;Maelzel adds speaking machine «check»		
Carl Gottlieb von Windisch	1783	*Lettes on Kempelen's Chess Player (Inanimate Reason)*	vK's friend and promoter	
Joseph Friedrich Rackneitz	1789	Poe's main source		
Boy Babbage meets Merlin	1800 c			
Johann Nepomuk Maelzel	1809 c 1815	Aquires and restores Turk Reaquires Turk	Bought from vK's son	
Babbage plays Turk	1820	plays and loses	(saw earlier in 1819)	
Robert Willis	1821	*An Attempt to Analyse the Automaton Chess*	Poe's closest source incl. tautology	

			Player	
Babbage	1822		Letter to Sir Humphry Davy (Royal Society)	
Maelzel with Turk in NY	1826		Feb. 3 arrives NY Harbor	Debuts April 13 NY; Philly base; 2 boys in Baltimore
American Chess Player	1827		Walker bros. in NY	
Whist Player			NY	
Brewster, Sir David	1832		*Letters on Natural Magic*	Harper's Stereotyped Ed. NY
Babbage	1834		Analytical Engine	
Locke, Richard Adams	August 1835		Great Moon Hoax	NY *Sun*
Poe sees Turk in Richmond	1835		December	
Poe	April 1836		"Maelzel's Chess-Player"	*Southern Literary Messenger* (derivative of Willis, Brewster)
Shclumberger (automaton) dies	1838		April	
Maelzel dies abord ship	1838		July 21 enroute to Philly from Havanna	
John Kearsley Mitchell	1838		Club to buy and analyze Turk	
Walker, George	1939		"Anatomy of the Chess Automaton"(actual method)	
Babbage(Lovelace/ Menabrea)	1843		Translation and Notes	

George Walker	1850	*Chess and Chess-Players*		
George Boole	1854	*An Investigation of the Laws of Thought* as mathematics		
Turk burns in storage	1854	July 5th at the Chinese Museum in Philly		
Melville	1855	"The Bell-Tower"	*Putnum's Weekly*	
Melville	1856	*Piazza Tales* ("The Bell-Tower»)		
Allen, George	1859	"The History of the Automaton Chess-Player in America»	Accurate details of solution and final days	
Babbage	1862	Difference Engine at London Exhibition		
Ellis, Edward S	August 1868	*The Steam Man of the Prairies*		
Hale, Edward Everett	1869	*The Brick Moon*	*Atlantic Monthly*	
Babbage	1871	dies		
Mitchell	March, 1874	"The Tachypomp"[an droid does math and verse]	*Scribner's Monthly*	
Enton, Harry (Cohen, Harry)	1876	*Frank Reade and His Steam Man of the Plains*		
Mitchell	May 4, 1879	"The Ablest Man in the World"	NY *Sun*	
Senarens, Luis P.	1879	takes over Frank Reade stories		
Stimson,	Nov.	"Dr.	*Scribner's*	

Frederic Jesup	1890	Materialismus"		
Bierce, Ambrose	1893	"Moxon's Master»		
Serviss, Garrett P.	1898	"Edison's Conquest of Mars"(*NY Evening Journal*)		
Leonardo Torres y Quevedo	1910	endgame automatic chess-player		
Karl Capek	1921	*R.U.R. (Rossim's Universal Robots)*		
Wiener	1948	*Cybernetics: or, Control and Communication in the Animal and the Machine*		
John Bates	1949-58	The Ratio Club		
Wiener	1950	*The Human Use of Human Beings: Cybernetics and Society*		
Turing	1950	"Computing Machinery and Intelligence" (Turing Test)		
Shannon and McCarthy	1955	*Automata Studies*		
Wiener	1964	*God and Golem*		
John Gaughan	1971	reconstructs Turk		

[1] See Brewster, Cohen, McCorduck, Rosenfield, and Wood for broad historical overviews of automata and thinking-machines.

[2] That speech and chess appear in multiple iterations of the thinking-machine along with mathematics, logic, and reason is a function of their close association with thought and/as contingency.

[3] It is beyond the scope of this project to trace the many and intricate ways in which the two were connected by acquaintance, common friends and colleagues, and common and purpose, but as McCorduck suggests, Wiener and Turing, along with von Neumann and other important researchers in the field, such as Shannon, Pitts, and McCulloch, "were all connected by friendship, by proximity, [and] by their fascination" with the new theories and technologies of thinking-machines (McCorduck 83).

[4] "I might just as well say 'I am walking, therefore I am a walk.' M. Descartes is identifying the thing which understands with intellection, which is an act of that which understands. [...] We cannot conceive of jumping without a jumper, of knowing without a knower, of thinking without a thinker. [. . .] So it seems that the correct inference is that the thinking thing is material rather than immaterial" (Descartes *Meditations* 122). Ironically, though Hobbes critiqued Descartes specifically for his characterization of thought as *non*material, Descartes' reception in England suffered from a perception that his metaphysics was too much like Hobbes' own materialism (Jessef 201).

[5] For example, Henry More's "Spirit of Nature . . . pervading the whole matter of the Universe, and exercising a Plastical power therein . . . directing the parts of the matter and their Motion" follows from his critique that the "Mechanical part" of the "nullibilist" Descartes' philosophy "has by its own nature so enticed scholiasts and half-educated men that all the phenomena of the World can be explained from local motion and matter alone" (qtd. in Jessef 204-205).

[6] It is for this reason that Norbert Wiener can claim in 1949 that questions framed around the vitalist/mechanist binary are "badly posed" (*Cybernetics* 44).

[7] Baker and Morris offer rationality, freedom, moral agency, and communicability as the conceptual links with thought, each of which can be read through Descartes as contingency (Baker and Morris 6).

[8] Descartes' actual position on substance dualism is, as has been noted, widely contested. Taking a somewhat different perspective than Gaukroger, Alanen suggests that this language "indicates the difficulty Descartes has in trying to formulate his view in the traditional substance-attribute terminology, which his own brand of dualism seems to undermine. To read him as having an ontological two-substance view" she continues, "is to take essential forms and substances in a traditional way [but] a better way (suggested by Descartes himself) is to read his mind-body distinction as a merely conceptual distinction" (71).

Alanen is certainly right that Descartes is struggling to grasp a conceptual distinction through an inadequate language of substance, but this discursive limitation is not merely a matter of not finding the right words. It is, rather, a matter of not finding the adequate concepts, and the language of substance which persists in both Descartes' writing and thinking *does* tend to give his dualism ontological force. It is perhaps as fair to say, then, that Descartes' language of substantial separation undercuts the status of his dualism as "merely conceptual."

[9] This shifting of associated images and concepts into new formations and relations through multiple iterations is precisely the dynamic expressed by the Bernoulli Shift, also known as the baker's transformation (Pepperell 58).

[10] Gaukroger argues that Descartes' early work shows a preoccupation with mechanism, and little interest in metaphysics, but that his late adoption of substance dualism, dating to shortly after the condemnation of Galileo, was precisely an attempt to legitimate and protect scientific research on the observable material universe by setting aside the soul, and, thereby, any perceived threat to religious authority. Thus, Gaukroger argues, it was of paramount importance for Descartes, on the one hand, to restrict the soul to the smallest domain possible, and, on the other, to defend that reduced domain of the soul against the encroachment of mechanism at

all costs (12-13; 292; *passim*).

Baker and Morris also "argue for the view that his fundamental project was to extend the range of phenomena which can be explained by the principles of mechanics to include the vegetative, sentient, and locomotive functions of living organisms," leaving only the rational soul as the uniquely human domain (8).

[11] Ultimately, it is the need to conceive of contingency and determinacy as mutually implicated functions that requires the emergence of these concepts.

[12] Descartes' concept of reason, of course, is part of what is at stake.

[13] Alanen makes a related point when she stresses the "real union" of mind and body, and argues that in Descartes' dualism mind and body are merely "conceptually independent" (47). Whether or not Descartes was aware of this in his writing is hard to say, but Alanen is right that his argument seems to function this way. In some sense, then, and quite ironically, the myth of Cartesian dualism that followed Descartes should be understood as a reaction against the mechanistic and monistic inclinations of his metaphysics.

[14] Discussing the role of Mersenne's mechanist philosophy in the development of Descartes' metaphysics, Gaukroger stresses the "inertness of matter" characteristic of both Mersenne's mechanism and mechanism in general during the seventeenth century. "When Descartes comes to investigate the metaphysical foundations of mechanism in his work [. . .] his programme will be in many ways a detailed and sophisticated development of Mersenne's" (149-150).

[15] Thus, both the vegetative and sensitive souls of Scholasticism "can be dealt with in mechanistic terms," and "Descartes' aim was to show that a number of psycho-physiological functions [. . .] could be accounted for in a way that did not render matter sentient" (Gaukroger 270, 278).

[16] "I am now dissecting the heads of various animals, so that I can explain what imagination, memory, etc. consist

in" (Descartes *Correspondence* 40). Baker and Morris observe that Descartes' program of "disambiguation" for terms like memory and imagination when referring to the distinction between the vegetative/sensitive functions and the rational soul, have lead to the very confusion he was attempting to ameliorate (32-33).

[17] Note that Descartes rejects Harvey's characterization of the heart as a pump, which could undermine his commitment to the inanimacy of matter (Gaukroger 271).

[18] See fn22 below.

[19] It is notable that Descartes apparently lived much of the time between the summer of 1614 and the fall of 1615 at Saint-Germain-en-Lay, where he would have visited the grottoes and fountains of the Royal Gardens, which featured hydraulically-powered statues that danced, played, and spoke (Gaukroger 62-63).

[20] This, of course, is precisely the discursive field explored by Melville in "The Bell Tower."

[21] The historical link between clockworks and automatons is well-known. Indeed, the earliest modern automata were essentially parts of elaborate cathedral depicting the phases of the moon, or Madonna and child, and, somewhat later, city clocks, featuring mechanical knights, trumpeters, and animals. Moreover, the smaller moving pictures, table ornaments, and other small toys that proliferated during the eighteenth century were often made by clockmakers (Standage 2-5).

[22] This trope of mechanical determinism, coupled with its correlative "perfection" and superior performance when compared to human beings, will provide the basis for one of the uncanny aspects of the thinking machine as it appears in many nineteenth-century literary texts.

[23] Descartes (not uniquely, but somewhat misleadingly) uses the term to indicate the "finer parts of the blood" (Descartes *Treatise on Man* 100).

[24] For rhetorical (and no doubt political) purposes, Descartes is not describing human beings *per se* in this text, but

"fictional men . . . intended to cast light on the nature of real men" (Descartes *Treatise on Man* 100-101, fn 1).

[25] Cottingham provides a note indicating that the last clause of this sentence appears in the 1647 French translation by Louis-Charles d'Albert, Duc de Luynes, approved by Descartes. It is also the language that La Mettrie will pick up in *Machine Man*.

[26] Baker and Morris warn against too facile an identification between consciousness and thinking when considering Descartes' use of the images of the clock and the machine. Arguing that "the Legend" takes too much function from the body, and therefore misses its non-thinking capacity for faculties like memory and imagination, they insist that "it needs to show that there is a contradiction in talking about a 'conscious machine'" (37).

Their argument, however, largely ignores issues of determinacy and contingency, and concerns, rather, a definition of the "conscious" that includes the functions traditionally ascribed to the vegetative and sensitive souls, and which, as we have seen, Descartes takes pains to ascribe to mechanism. This sort of slippage does certainly occur in nineteenth-century discussions surrounding the Difference Engine (including Babbage's own writing), so that, despite resistance to the imagery, there need not be anything anti-Cartesian or non-dualist about a machine with a memory; however, as Baker and Morris point out, such faculties by then were fixed firmly in the realm of the soul within the popular imagination.

Their discussion of Descartes' distinction between clockwork and machines is more dubious, however, and the example they give seems to contradict their point that Descartes uses clockwork and machines to signify different things.

> We see clocks, artificial fountains, mills, and other such machines which, although only man-made, have the power to move of their own accord in many different ways. But I am supposing this machine *to be made by the hands of God*, and so I think you may reasonably think it capable of a greater variety of movements (CSM I: 99; qtd. in Baker and Morris, emphasis theirs).

But, surely, here Descartes explicitly *names* the clocks and mills as machines; the body-machine is not opposed to clockwork, as they argue, but a particularly fine example of it.

Furthermore, in the other passage they cite, Descartes refers to "automatons, or moving machines" (*Discourse* 139), with "moving machines" serving as the definition for "automaton"-- and, as noted above, automatons are most certainly clockwork.

[27] This is why the conceivability of the thinking machine in forms like Babbage's calculating engines or Maelzel's chess-player constitute a threat to the concept of the human as understood through Descartes, or anti-/Cartesianism. That is, if thought is no longer a strictly human function, there is no remaining uniquely human function or quality through which to identify the category. Put differently, if the only difference between humans and machines is the capacity for thought, then the conceivability of the thinking-machine eliminates the difference between these conceptual categories.

[28] In the *Meditations*, for example, Descartes concludes that "it is certain that I am really distinct from my body, and can exist without it" (54). The authorized French translation includes a footnote which elaborates that the I is "my soul, by which I am what I am" (Descartes *Meditations on First Philosophy* 54).

[29] Descartes' emphasis on the *rules* of reason somewhat complicate this easy identification of thought with contingency (e.g. the *Discourse*, and the *Regulae*, which explicitly links reason to mathematics, the language of a mechanical universe). These texts suggest that contingent thinking is, after all, for Descartes, the product of deterministic rules of reasoning, and show that already, within Descartes, there is the possibility of contingent function emerging from determined rules of play. In this sense, Descartes' rationalism is itself the image of a thinking-machine.

[30] Alanen rightly reads Descartes' use of the words for soul, mind, and thought as "different names of the same thing," namely self-consciousness or awareness (Alanen 78). The point for her, is that Descartes' concept of thought encompasses those functions which (for Descartes) cannot be explained in purely material/mechanical terms, and belongs to a conceptual rather than a metaphysical order (Alanen 79). Again, however, I think it is more accurate to say that Descartes is struggling to articulate a conceptual category within an inherited metaphysical

discourse, so that within his own discourse, the language of function tends to dominate and overwhelm his description of substance. But, this is precisely why the component (functional) concepts with which he identifies his various concepts of the human, animal, machine, body, and soul, are so easily dislocated from those concepts and available for redistribution.

[31] And, to the extent that this formulation grounds nineteenth-century America's understanding of the human, both Babbage and Darwin represent crises in the concept of the human.

[32] While the formula as stated is not logically reversible, Descartes would also say that actions depending on reason must be reckoned human.

[33] Descartes' interest in automata extended to a proposal in his personal notes for a man-machine operated by magnets, as well as for a flying pigeon and a spaniel hunting a pheasant (Rosenfield 4).

[34] Given a metaphysics of substance dualism, or even a weaker conceptual dualism, a speaking automaton is indeed--by definitions--inconceivable. It is easy to see, however, as did many of Descartes' contemporaries that unless one begins by assuming separate substances, there is nothing to prevent the attribution of the function of speech to mechanism. It is, of course, this weakness in Descartes' argument that La Mettrie will exploit to great effect.

[35] Interestingly, the Baron Von Kempelen, most famous for his fraudulent chess automaton, spent much of his life perfecting hand-operated speech synthesizers for the mute. Though Kempelen's speaking machines (like his chess-player) were guided by a human operator, and were not, therefore, thinking-machines, it is nonetheless true that they "emulated a uniquely human capability, never before witnessed in a machine" (Standage 76-81).

Strangely ignoring the long history of mutual association between chess-playing- and speaking-machines, Alanen writes, "thinking is not, any more than meaningful speech, like a game of chess, the rules of which determine every admissible move and must be mastered in order to play" (105). But, this is a very interesting thing to say, since, as we shall see, it is precisely the play between rules of the game and contingency that leads Poe to *identify* thought with chess, and thus to rule out the conceivability of a mechanical chess-player. Presumably, it is the presently undeniable fact that machines *do* play chess in 2003 that leads Alanen to have to deny the function of chess-playing as a test of intelligence--at least as long as she wishes to maintain her stance against the conceivability of the thinking-machine.

[36] This premise will serve as Poe's principle argument in his 1837 (check date) exposé of Maelzel's fraudulent chess-playing automaton.

[37] In this same passage, Descartes concludes that the souls of animals must be "completely different in nature from ours" (Descartes *Discourse* 140). As I have already noted, in most other instances, Descartes seems to insist more strongly that the soul *is* thought itself, and that therefore animals lack souls entirely.

[38] John Haugeland has made a similar observation, suggesting that "Descartes is saying that machines can't think (or talk sensibly--he's anticipated Turing's test too) *because* they can't manipulate symbols rationally" (Haugeland 36). But, Descartes does more than simply anticipate Turing--he puts into play the analytical rules and the conceptual pieces (that is, the formal system) which will organize both Turing's "imitation game" and his chess-playing program (both of which are run on universal Turing machines). Thus, Alanen, somewhat more insightfully, suggests that if Haugeleand is right, we may well even credit Descartes for "having inspired--in spite of himself" the idea that a machine could pass the test for thought (Alanen 96).

There is something strange, however, about Alanen's reading of Haugeland, whose argument she casually dismisses. Apparently confusing the *Turing test*, which applies a standard of contingent functioning within linguistic rules to the problem of machine intelligence, with the *Turing machine*, which represents a simple set of predetermined rules for processing data on paper, she objects that the Turing machine "would not count as thinking in Descartes' sense" (Alanen 96).

But, this precisely misses the point. While every Turing machine is essentially the calculation of a specific algorithm (or deterministic functional rules of play), contingency can be designed into any program by layering any number of such algorithms and including conditional branching steps (Haugeland 47-84). In fact, Turing suggested his famous test at the very same time that he was theorizing a computing machine built as a trainable (that is, contingently functioning) network of randomly arranged neuron-like elements--*and, these can be run on Turing machines* (Copeland 403; Turing "Intelligent Machinery" 465-475).

While Alanen is certainly right that Descartes flatly rejected the possibility of contingent thought as a process of deterministic bodies, it is equally the case that Descartes and Cartesianism put into play the conceptual components from which this possibility emerges. In other words, it is not Descartes' concept of thinking that is at stake so much as the multiple transformations to which it is open. Thus, the iterative work of Cartesianism (broadly understood) has been precisely to show that deterministic rules of play at one level can produce contingent function at other, higher levels of organization. Whether this undermines or reifies Descartes' concept of a thinking-thing is a matter of perspective.

[39] Descartes chose not to publish during his lifetime the text of which both *The World* and *Treatise on Man* were parts, after learning of Church's condemnation of Galileo (Descartes *Treatise On Man* 79; *Correspondence* 41).

[40] This point echoes Hobbes' objections to the *Meditations*. The fact that the discursive formulation of La Mettrie's materialism appears upon its face to be indistinguishable from Descartes' description of the soul is significant for understanding the way in which the concepts of cybernetics and the posthuman can be said to emerge from the multiple iterations of Descartes' dualism.

[41] "[W]hy divide into two what is obviously only one?" (La Mettrie *Machine Man* 30).

[42] It is not at all clear that La Mettrie allows for any contingency in his universe whatsoever. Certainly, his concept

of the machine seems to be identified by determinacy, and this is clear in the discussions of clockwork. Insofar as it is a machine, then, the human body (which includes the mind) is also seen to be determinate. That is, our functioning "depends on the way our machine is constructed" (La Mettrie *Machine Man* 8), and the nature and the actions of the human being are *as deterministic* as those of any animal or machine--that is, absolutely. (This is why he counsels compassion for criminals and idiots). However, La Mettrie's discussions of the imagination and of animal speech, having cast it out of the soul, seem rather to relocate contingency within the material.

[43] However, for La Mettrie there is a recursive relationship between thought and speech, with language organizing and shaping thought through "arbitrary signs" (La Mettrie *Machine Man* 13), so that between language and thought, "the one helped the other" (14).

[44] The story is recounted in many sources, including Gaukroger, Wood, and Rosenfield (who quotes Anatole France). Gaukroger, who notes that the story circulated widely during the nineteenth and early twentieth centuries, can find no evidence of the story before the eighteenth century, and believes it to have originated as a tool of propaganda against La Mettrie's Cartesian-inspired materialism (Gaukroger 1-2). Anyone who has read even a fair amount of nineteenth-century American science fiction knows that the villainous scientists are nearly always materialists.

[45] The story also bears a striking resemblance to Mitchell's 1879 short story, "The Ablest Man in the World," though whether Descartes' influence on Mitchell was a direct one, or more viral, is unknown.

[46] See Chapter 3.

[47] Recall that Descartes, too, understood mathematics to be the formal language of reason.

[48] Babbage's favorite term for the running of any specific calculating program.

[49] Principally, from two 1822 articles appearing in *Brewster's Journal of Science* and the *Memoirs of the Astronomical Society*, and his 1832 volume on *The Economy of Manufactures and Machinery*.

The latter was reprinted widely in the United States, as well as in Europe (Morrison and Morrison 374).

[50] Poe tells us that his own source on the calculating engine and other automata is Brewster's *Letters on Natural Magic* ("Maelzel's Chess-Player" 346).

[51] Such as in this short newspaper article from 1834: "but for which they might claim a place on the page of fame so brilliantly illuminated that the names of Fulton, Bolton, Watts, Sir Humphrey Davy, and Mr. Babbage would be blackened into obscurity" ("Perpetual Motion").

[52] "The operation of Mr. Webster's pen is somewhat like the process of Babbage's machinery for the mathematics: there is an abstraction of all party and personal influences; questions are settled by simple reason and logic alone" ("Mr. Webster's Report"). See also the previous footnote.

[53] Doron Swade makes this same case, tracing the recursive relationships among Babbage's, Lardner's, Nicola's and Brewster's published accounts of the calculating engines, pointing out that,
> All of these accounts, without exception, are directly based [on] sources provided to the authors by Babbage. [. . .] the voice in each of these accounts is unmistakably Babbage's. [. . .] The issue of authorship is further blurred by Babbage publishing two of the accounts in full in his own published works [and] the historical cannon is effectively dominated by Babbage's own account" ("Shocking Truth").

[54] As to insights into the adult Babbage's psyche, he begins his Preface to *Passages* by noting that in reply to numerous calls for an autobiography, he has sent a list of his published works, "with the remark that they formed the best life of an author" (vii). "I have no desire to write my own biography," Babbage asserts. "This volume does not aspire to the name of an autobiography" (*Passages* vii). Whereupon he embarks on the first of four full chapters tracing his personal origins, from prehistoric tool-makers, through the earliest Babbages, to his own childhood, youth, and early days at Cambridge.

I am inclined to set aside the obvious charges of self-absorption and egotism, however, and point out that the narrative framing of *Passages* is consistent with a more humble view that understands a life and its work, as well as its daily minutia and mental life or subjectivity, as a complex-system communicating with and embedded within other complex-systems. Thus, in some sense, the narrative structure and content of *Passages* can be understood as an iteration of the theories Babbage explores in *The Ninth Bridgewater Treatise*. The plans and explanations for the Analytical Engine would be another iteration in this pattern.

[55] We can see in this formulation a tension within Descartes that Babbage will find increasingly difficult to overlook. For, if the mind and reason are the contingent functions of the human being, then it is a prima fascia paradox to speak of the laws that govern the mind. Babbage's reflections on this problem in *The Ninth Bridgewater Treatise* are among his most interesting.

[56] The experiment was deemed "inconclusive" because, while no devil appeared, the young Babbage was struck the same night by an inability to remember the familiar Lord's Prayer. This, he worried, was God's punishment for his religious doubt (Babbage *Passages* 9).

[57] In the Preface to the Second Edition of *The Ninth Bridgewater Treatise*, Babbage makes the same case, defending "the truths of Natural Religion [which] rest on foundations far stronger than those of any human testimony" (xv). But, [r]easoning," Babbage counters, "is to be combated and refuted by reasoning alone" (xiii).

[58] For Babbage, however, the principle proves fatal rather than instrumental to his faith in God.

[59] Descartes not only used the automaton as a trope in his discussion of animal and human bodies, but collected and experimented with them, as well (Wood 4).

[60] Again, Babbage's desire to locate the catalysts of large formations in his later life within small childhood incidents is consistent with his theoretical musings on the relationship between determinism and contingency in *The Ninth Bridgewater Treatise* and elsewhere, as well as with later complexity theories

and philosophies of becoming. In short, Babbage is implicitly arguing that these minor events in his early life served as the first iterations of the governing patterns through which his later life would emerge.

[61] Babbage tells us that he attended several such exhibitions with his mother (Babbage *Passages* 12).

[62] Interestingly, Babbage often uses a language of seduction when writing of the Silver Lady. He tells us twice that she is "uncovered" when he first sees her, describes her eyes as "irresistible" (suggesting her humanity), and makes a point of describing the stylish clothing he gives her as verging on immodesty; he almost always refers to the automaton as female, writing of her charms and grace, and generally treating her as a "lady." However, in describing the disassembly that he preformed upon acquiring her, Babbage writes primarily of the "automaton" and its "mechanism" (Babbage *Passages* 12, 273).

[63] Thus, it is no accident that Babbage comes to think of nearly every one of his engineering solutions as essentially an innovation in a system of signs.

[64] Thus, when Poe writes that Babbage's calculating engine, as a "pure machine," and in contrast to the counterfeit automaton chess player, "proceeds [. . .] to its final determination, by a succession of unerring steps. [. . .] finite and determinate" on principles of mathematical determinacy (Poe 349), he both echoes Descartes' discussion of mechanism and determinacy, and fails to understand one of Babbage's principal technological and philosophical insights.

For Babbage, it is in part his ability to understand deterministic functions as linguistic (that is, as systems of signs expressing contingencies through deterministic rules of operation) that helps him to begin to collapse the Cartesian distinction between determinism and contingency in a way that resonates with twentieth-century complexity sciences and the communication-as-control principle of Norbert Wiener (Wiener *Cybernetics* 39).

[65] Organized as the Analytical Society, this group included John Herschel, son of the famous astronomer Sir William Herschel, who would become an important and popularly-

known astronomer in his own right; and George Peacock, who would later spearhead the reform of British mathematics as a dean at Cambridge (Hyman *Pioneer* 25).

[66] Though, clearly, in the case of the calculating machines, these problems became more general and theoretically implicated toward the end of his life.

[67] Such tables were also widely used by insurance actuaries, bankers, and manufacturers, as well as in the sciences. Astronomical tables used to chart navigation were of particular concern, since even a few small errors could, and did, sink ships. Indeed, Britain's reliance on such tables by the early nineteenth century cannot be overestimated, and the need to eliminate human error from both the calculating and the printing processes drove not only Babbage's first efforts at mechanical invention, but also the Royal Society's willingness to fund them (Swade 12-13, 32).

[68] Again, following Babbage's own narrative structure, we can think of each of these items as an iteration of an emergent pattern which recursively describes the complex-system that could be called "Babbage," or "the life of Babbage."

[69] The term referred at that time to human beings who performed and recorded mathematical operations.

[70] Hyman cites the account as "C. B. *History of the Invention of the Calculating Engines*, 10, Buxton papers." Both Swade, and Morrison and Morrison, citing the account given in Buxton's 1935 article, quote Babbage as saying, "I wish to God these calculation had been executed by steam." Morrison and Morrison, quoting the same source, report Herschel's reply as, "It is quite possible" (Swade 10; Morrison and Morrison xiv).

[71] The selection on Babbage from Brewster's *Natural Magic* often appeared unattributed in American newspapers and magazines under general subject headings. These included "Miscellany," in the *Salem Gazette* in November 1832 and the "Mechanics' Department" in *New York Weekly Messenger* in 1833. It is likely that the article appeared at one time or another in most major markets. Lardner is cited or quoted frequently, as well.

[72] Lardner's extensive account of the errata in existing tables is not only heavily influenced by his correspondence with Babbage, but recursively becomes Babbage's primary source when he turns in his autobiography to his own account of the problem of errata (Babbage *Passages* 103).

[73] While Brewster was widely reprinted in the U.S., so were other London-based reports of Babbage. Thus, already in 1822, apparently picking up the link between the human and error directly from Babbage's open letter to Sir Humphrey Davy, the Hartford *Times* reported that with Babbage's machines, "calculation may be performed with a facility and accuracy hitherto unknown [and enable] printing mathematical tables without the aid of a compositor." The entire article from *New Monthly Magazine* was also reprinted in the *Providence Gazette* from the *Boston Daily Advertiser*.

[74] Consider, for instance, the way in which our contemporary concept of "calculation" includes determinism as a component concept. Furthermore, a "calculating" person acts without warmth or sympathetic emotional motivation, but only on the basis of "cold," "hard" fact or personal reward. In this reversal of the Cartesian schema, reason ceases to signify the human, but comes to signify the automaton instead.

[75] Both Leibniz and Pascal were widely known for their calculating machines, and are frequently mentioned in the newspaper accounts of Babbage's Difference Engine in the United States.

[76] Written almost thirty years later, Babbage's language here resonates strongly with Poe's discussion of the "pure machine" in "Maelzel's Chess Player" (349-350).

[77] These examples from *Passages* were reprinted in the United States in an unattributed 1864 *Harper's* review of the book ("Recreations of a Philosopher").

[78] Again, for Descartes, a thinking-machine is "not conceivable" because contingency is the central component in his concept of the human, and it is this concept of contingency which bridges his concept of the human with that of reason.

[79]

[T]he public having erroneously imagined, that the sums of money paid to the workmen for the construction of the engine, were the remuneration of my own services, for inventing and directing its progress; and a Committee of the House of Commons having incidentally led the public to believe that a sum of money was voted to me for that purpose,--I think it right to give to that report the most direct and unequivocal contradiction (Babbage *Bridgewater* 188).

[80] This was a serious problem, since the standardization of manufacturing equipment was yet to come (in large part due to innovations and later campaigning by Clement's journeyman on the Difference Engine project, Joseph Whitworth), and Clement owned the lathes and other machinery which he had designed in order to produce the parts for the Engine. Clement also had in his physical possession those parts, as well as mechanical drawings and other irreplaceable items. The parts of the unfinished machine, along with drawings, were eventually returned to Babbage in 1834, but the manufacturing equipment remained in the hands of Clement, as per the custom of the time (Swade 46-47; 60-61; 66-71).

[81] Babbage includes in his autobiography a reprint of a statement by Sir Harris Nicolas, originally circulated as an anonymous pamphlet in 1843, setting out Babbage's version of events (52-74). This was the same year that the incomplete first Difference Engine was turned over by Babbage to the Museum of King's College in London, along with any claim to rights of ownership. Though incomplete, a fully working section capable of producing calculations had been assembled in 1832. This small calculating engine was given a public audience at the 1862 International Exhibition in London (Babbage *Passages* 113). Much to Babbage's consternation, a duplicate of a much-celebrated Swedish difference engine, of a different design but inspired by his own, was commissioned by the British government a few years later, and did eventually go to work calculating tables for the British government (Babbage Passages 83).

[82] In 1991, in an effort to prove that Babbage *could* have built a

working Engine using early nineteenth-century technology, had he had the necessary backing, a team at the Science Museum in London, under the direction of Doron Swade, completed a fully functioning version of the second Difference Engine based closely on Babbage's drawings (Swade 223-224, 283; Science Museum). Numerous amateurs have also been inspired to build calculators based on Babbage's plans, including several by Babbage's son Henry Prevost Babbage (Swade 314-316), Andy Carol's Difference Engine Number Two, built entirely of Legos (Carol), and an Analytical Engine built with Maccano pieces (similar to Erector) by Tim Robinson (Robinson).

[83] That is, a de-cision, an actualization of the virtual, a literal "cutting-down" of the multiple into one line of becoming. In any moment of de-cision, contingency offers multiple becomings, while each choice taken both closes lines of becoming and opens an infinity of new de-cisions. This functional ability to process contingency, for Descartes, was the power of mind, (free) will, and the essence of the human; for Babbage, it was the power of mind as mechanism.

[84] "[T]he Difference Engine is not intended to answer special questions. Its object is to calculate and print a *series* of results formed according to given laws" (Babbage *Passages* 36). That is, in order for the machine to produce a series of specific results, some contingency must be built into its general laws of operation.

[85]
> The illustration which I shall here employ will be derived from the results afforded by the Calculating Engine.[. . .] and this I am the more disposed to use, because my own views respecting the extent of the laws of Nature were greatly enlarged by considering it (Babbage *Bridgewater* 32-33).

[86] Descartes does not always appear to be clear or coherent in his various discussions of memory, sometimes seeming to treat it as a function of thought ("There are various modes of thought such as understanding, imagination, memory, volition, and so on" [Descartes *Correspondence* 216]), and other times as a function of the body (as when he writes of "dissecting

the heads of various animals, so that [he] can explain what imagination, memory, etc. consist in" [Descartes *Correspondence* 40]). However, Descartes quite coherently reconciles these apparent discrepancies in the Fifth Reply, when he writes that,

> So long as the mind is joined to the body, then in order for it to remember thoughts which it had in the past, it is necessary for some traces of them to be imprinted on the brain; it is by turning to these, or applying itself to them, that the mind remembers (Descartes *Meditations* 247).

Thus, in addressing the problem of memory in the human mind, and while again discursively linking substance with function, Descartes describes precisely the problem that addresses Babbage with respect to "carriages" in the first Difference Engine.

[87]
> [The Analytical Engine] will calculate the numerical value of any algebraic function—[. . .] at any period previously fixed upon, or contingent on certain events, it will cease to tabulate that algebraic function, and commence the calculation of a different one, and [. . .] these changes may be repeated to any extent.
> The former engine could employ about 120 figures in its calculations; the present machine is intended to compute with about 4,000 (Babbage *Bridgewater* 187).

[88] This arrangement provided what Wiener would call "feedback," and begins to suggest a recursive relation between language and thinking that has implications for Turing's Test.

[89] "When we attempt to perform such additions by machinery we might follow exactly the usual process of the human mind" (Babbage *Passages* 43).

[90] "Not only does this system of wheels calculate, as though it were a living and reasoning thing, but even writes down and prints off its labor" ("The World's Progress").

> "A proposition to reduce arithmetic to the dominion of mechanism--to substitute an automaton for a compositor--to throw the powers of thought into wheelwork. [. . .] the process of reasoning is performed by inanimate matter" (Lardner 164; 166).

[91] Babbage's Deist tendencies surfaced early in his career at

Cambridge when he published, with Herschel, *The Principles of pure D-ism in opposition to the Dot-age of the University*. While the main point of the text was to protest the British university system's exclusive use of Newton's notation over that of Leibniz, the subtext pointed to a religious controversy in which, for Babbage and Herschel, Deism as well as D-ism represented the sensible approach (Babbage *Passages* 21).

[92] It is not likely that Babbage took the idea of an intentional Creator seriously. Babbage "went down" at Cambridge for attempting to defend the thesis that God was material, and most of the *Bridgewater* language regarding the First Cause is quite impersonal. Furthermore, Babbage's example of a mechanism "susceptible of having embodied in its mechanical structure, that more general law of which all the observed laws were but isolated portions" (Babbage *Bridgewater* 40), suggests that for him, God functioned as a trope for Natural Law unfolding as the universe.

[93] Babbage is expressly addressing Hume's refutation of miracles here. He rejects Hume's appeal to probability and natural law as arguments against belief in miracles, and calculating the probability of reliable independent witnesses agreeing in error, argues that miracles are *only* plausible, and indeed highly probable, as the unpredictable but necessary consequences of natural laws not yet fully understood (Babbage *Bridgewater* 124-131, 149-158).

[94] That is, "general."

[95] Babbage devotes a section of his autobiography to answering various metaphysical questions, including the question of whether or not animals think. His conclusion is that we cannot possibly know (Babbage *Passages* 304).

[96] One of Babbage's most prized possessions was a portrait of Jacquard woven in silk (Babbage *Passages* 127, 228-229).

[97] Babbage's use of the word "developed" here is interesting, as it tends to suggest a potential openness to these operations not imagined for the Difference Engines. This is the point of departure for Gibson and Sterling's *The Difference Engine*, an alternate-history novel in which Babbage's Analytical Engine is not only built, but becomes the pervasive infrastructure

for a nineteenth-century British Empire built on information technology. The introduction of a recursive, "infinitely iterated," contingency into the system through a stack of cards written by Ada Lovelace unleashes a process through which London emerges as a newly sentient being with the Analytical Engine as its brain (Gibson and Sterling 421-422, 428-429).

[98] From an 1852 letter to Lord Derby seeking a resolution to the government's involvement with the first Difference Engine.

[99] That is, usual with regard to the practices of human calculators.

[100] Lovelace's translation of Menabrea's detailed and technical 1842 discussion of the Analytical Engine, along with her own extensive notes, running to three times the length of the original text, was published in London in 1849. Babbage worked closely with both friends (Hyman *Selected Works* 243; *Pioneer* 183-185; Babbage *Passages* 102), and authorized their work, noting in particular Lovelace's grasp of "the very difficult and abstract questions connected with the subject," and confirming (presumably against chauvinistic doubts concerning a lady's mental capacity) that the mathematical examples represented her own work (*Passages* 102).

Babbage admits to having done the work on the Bernoulli numbers, "to save lady Lovelace the trouble," but adds that she was forced to send the work back to him with important corrections (*Passages* 102). Hyman dismisses "the quaint story of Ada as the world's first programmer" as "nonsense." As he suggests, Babbage clearly deserves this appellation. However, Hyman goes on to say that Lovelace "had a considerably better grasp of mathematics than some commentators have allowed" (Hyman *Selected Works* 243).

While I can find no record of the essay and notes having been reprinted under a U.S. imprint, they are referenced in an 1862 article in *Scientific American* report on the Great Exhibit reprinted from the London *Times*, and imported copies would presumably have circulated widely in the U.S. on the basis of Babbage's fame.

[101] Menabrea's discursive difficulties suggest Norbert Wiener's comment that "we were seriously hampered [. . .] by the absence

of any common terminology, or even of a single name for the field" (Wiener *Cybernetics* 11).

[102] That is, the material embodiment of mental operations.

[103] This language is highly resonant with Deleuze and Guattari's "between two," and suggests the philosophy of becoming that Babbage seems on the verge of grasping in *The Ninth Bridgewater Treatise*.

[104] "[I]t is for all practical purposes impossible for a machine to have enough different organs to make it act in all the contingencies of life in the way in which our reason makes us act" (Descartes *Discourse* 140).

"But from the first move in the game of chess no especial second move follows of necessity.[. . .] in proportion to the progress made in a game of chess, is the *uncertainty* of each ensuing move A few moves having been made, no step is certain" (Poe "Maelzel's Chess-Player" 350).

[105] Norbert Wiener's iteration of this problem is, not surprisingly, nearly identical to, though somewhat more complex than Babbage's (*Cybernetics* 165; *Human Use* 175). Interestingly, he cites Poe and von Neumann, but not Babbage.

[106] "The great object of all my enquiries has ever been to endeavor to ascertain those laws of thought by which man makes discoveries" (*Passages* 340).

[107] "[I]t is indeed of great importance that calculations made by machinery should not merely be exact, but that they should be done in a much shorter time than those performed by the human mind" (*Passages* 44).

[108] The Heisenberg Uncertainty Principle states that "The more precisely the position is determined, the less precisely the momentum is known" (Heisenberg); or, that for any pair of conjugate variables, the more that is known about the qualities of one term, the less can be known about the qualities of the other.

[109] This analysis also suggests an Oedipal reading, as offered by Freud in his essay, "The Uncanny" ("Das Unheimlich"); and, beginning with Shelley's *Frankenstein*, the Oedipal narrative does indeed pervade the literary treatments of the subject on

both sides of the Atlantic. However, my argument implies that rather than explaining the literature, this Oedipal narrative itself needs to be explained. More will be said about Freud's essay on "The Uncanny" in the next chapter.

[110] By 1790, over a half dozen pamphlets had been published on the Turk (Wood 64).

[111] Extensive accounts of the Turk's early career with von Kempelen, as well as its purchase by Maelzel and subsequent travel to England and the United States can be found in Allen, Walker, Carroll, and Standage. The most current and reliable of these is Standage, whose sources include the others, as well as material unavailable until recently. Thus, I have relied heavily on Standage for secondary historical material relating to the chess-player, particularly in Europe; and where Standage seems to have gathered material from these other secondary sources, I have generally cited Standage.

[112] So called because of its garb, presumably a reference to the "mysterious east," but also to the origins of the game.

[113] Standage writes that the "exchange of letters highlighted a theme that was to reappear repeatedly throughout the Turk's career: the question of whether a machine could make unplanned movements under its own initiative" (Standage 35). As Wood suggests, "there was an urgency of argument and anxiety in the writings [...] connected with the question of what was human, and only human (Wood 64).

[114] Standage also stresses the way in which the Turk's "apparent ability to respond to the moves of its opponent [. . .] set it apart from previous automata," so that, whereas previous mechanisms (such as Vaucanson's duck or Babbage's dancer) "simply did the same thing over and over again," the Turk appeared "interactive" (Standage 35). In other words, previous automata had reproduced the deterministic functions of the body, while von Kempelen's automaton chess-player appeared to reproduce the contingent functions of the human mind--to iterate (the image of) the thinking-machine.

[115] Von Kempelen largely considered this to be an unwelcome

and trivial distraction from his more serious work, including the construction of a speaking machine to assist the disabled (Standage 36-42; 76-81). Unlike these impressive devices, the chess-player's genius was hidden within its internal mechanism, while its apparent function was illusory.

Indeed, von Kempelen never made any claim that the device was anything but an illusion, a *"bagatelle"* (Windisch, qtd. in Brewster 273; Wood 68), and, in fact, boasted even before building it that it would achieve a "deception" greater than those performed at Maria Theresa'a court by the French conjurer Pelletier (Standage 17-19). It seems, however, that von Kempelen enjoyed keeping alive the concept of the thinking-machine; for, oddly, though he always spoke of the chess-player in terms of effect and deception, he never revealed the secret of the device, which would surely have had the effect of highlighting his mechanical genius while simultaneously discouraging unwelcome interest in the spectacle of the chess-player.

[116] One has to wonder how responsible the Turk might have been for setting off this craze in the early 1770s, as the first accounts and analyses were circulating among the same social groups at just that time.

[117] Grimm in particular seems to have difficulty writing his objections to the chess-player within the Cartesianism that he assumes. While appearing to make an *a priori* argument against the conceivability of a thinking-machine, he writes that the machine could "not know." But *not knowing* is not so much the lack of capacity for knowing (or thinking) as it is a *contingent function of knowing*--that is, an error in thinking, and, as such, a function of a thinking mind.

[118] This is the explanation to which Babbage disparagingly refers in his autobiography (Babbage *Passages* 353).

[119] Edmund Cartwright, upon seeing what he apparently took to be a genuine chess-playing automaton in London in 1784, was inspired to patent the power loom (Standage 68-69). This is a point of considerable interest, since it was Jacquard's punch card system that subsequently inspired Babbage's designs for the Analytical Engine, as well as his plans for a *chess-playing*

machine.

[120] Standage also reports that Windisch was a friend of von Kempelen, and that the pamphlet "has von Kempelen's fingerprints all over it" (Standage 62).

[121] Charles Babbage's annotated copy of Windisch currently resides in the British Library (Standage 140).

[122] This did not refer to the human-operated speaking machine designed by Kempelen, but rather to an unrelated exhibition featuring what appeared to be an independently speaking automaton (Carroll 21). But, it is interesting that Thicknesse attacks both the chess-playing and the speaking machine in the same pamphlet, as both are important tropes linking the machine with human-like intelligence, and figure prominently in the history of the thinking-machine. And, both Kempelen and, later, Maelzel must have understood the connection between speech and chess, for the chess-player, as originally exhibited, routinely "spoke," by spelling out answers to questions from the audience (Carroll x), and, Maelzel, upon acquiring the chess-player, added a speech function, so that, as appropriate, the Turk announced, "check."

Furthermore, Kempelen's own speaking machine was quite successful and widely admired. As Standage observes, both "emulated a uniquely human capability, never before witnessed in a machine" (Standage 81). But, significantly, this is because both speech and chess require the kind of contingent functioning that *identifies* the human. More will be said about the relationship between speaking machines and chess-playing machines in Chapter 4.

[123] Though Poe's argument is strikingly like that of Racknitz on this and several other points, Wimsatt argues convincingly, through archival research, that Poe's sole sources were Brewster, and perhaps a story appearing in a Baltimore weekly paper (Wimsatt 146, passim). It is exceedingly clear when reading Brewster and Poe together that Brewster was Poe's main, if not only, source.

[124] The chess-player's first match under its new ownership was against Napolean Bonaparte (Standage 105-112).

[125] Charles Babbage saw the Turk in London in 1818, and again in 1819, when he played and lost a game to the chess-player (Standage 140).

[126] Wimsatt does an excellent job of sorting out Poe's written sources, and concludes that Poe owes "almost everything" of his solution to Brewster, whose own text is largely dependent on Willis (Wimsatt 47). The Baltimore article to which Poe refers is apparently *not* a reprint of Willis's pamphlet. But Poe does not, then, address the contents of the Baltimore article, but instead discusses Willis (without attribution, following Brewster), as reported by Brewster in *Letters on Natural Magic*, which Poe readily acknowledges as a source. That Poe takes his details from Brewster, who gets them from Willis is suggested by Poe's criticism of Brewster's assessment of the worth of Willis's argument (Wimsatt 145-147; Poe "Maelzel" 321-322). In addition to the *Letters on Natural Magic*, however, Poe may also have seen Brewster's review of Willis's pamphlet in the *Edinburgh Philosophical Journal*, which largely reprinted the essay (Wimsatt 145). This seems likely, since Willis' argument that

> [t]he phenomena of the chess player are inconsistent with the effects of mere mechanism [as] the movements which spring from it are necessarily limited and uniform. It cannot be made to usurp the faculties of mind; it cannot be made to vary its operations, so as to meet the ever-varying circumstances of a game of chess [or] usurp and exercise the faculties of the human mind"

is not reported in the *Letters* (qtd. in Standage 130; Wood 74).

Additionally, it is difficult to imagine that, given the attention that the chess-player received in the United States, Poe would not have seen any number of newspaper clippings, or overheard conversations, drawing from any number of earlier sources.

[127] It was apparently in Richmond, Virginia, in December 1835, that Poe saw the chess-player exhibited (Standage 176).

[128] The piece was written by the *Post*'s editor, William

Coleman, whose interest in chess was well-known in Europe, and apparently to Maelzel, who visited him within a few days of landing in New York (Allen 428; Carroll 69).

[129] The article was widely reprinted from *Noah's National Advocate*.

[130] Allen writes that, "the newspapers [. . .] were filled with detailed accounts, [. . .] communications that revealed his secret [. . .] and confutations" (430). A further measure of the automaton's impact and popularity within the United States is the proliferation of copycat automata, including a whist-playing device and at least two American Chess-Players, all of which should be understood less as exercises in philosophy than in entrepreneurialism. Maelzel, unable to purchase the first American Chess-Player (constructed and exhibited by the Walker brothers), eventually bought and destroyed the second American Chess-Player in order to prevent competition, but kept a Whist-Player from the same inventor (Allen 455-459; "M. Maelzel's Automata"; "Mr. Maelzel"; "More American Ingenuity"; "To the Lumbermen of New Hampshire").

[131] Maelzel toured until his death at sea in 1838. The chess-player was bought by John Kearsley Mitchell, reconstructed for private exhibition, and finally placed in the Chinese Museum in Philadelphia, where it was destroyed by fire in 1854 (Wood 80-81; Standage 188-191).

[132] It is remarkable that the writer here attributes to the chess-player the function of speech.

[133] This problematic actually involves both the mechanical and philosophical questions, and is exemplified by Robert Willis, whose pamphlet Poe accused of being "exceedingly unphilosophical," while largely adopting its contents: "the movements which spring from [mechanism] are necessarily limited and uniform" (qtd. in Standage 130). Variations of this argument would persist for decades, across England, Europe, and finally the United States, Poe's version being not only the best-known among them, but also the most urgently (and perhaps anxiously) argued.

The more general and philosophical category of questions regarding the conceivability of a thinking-machine, of course, begins with Descartes.

[134] See Wimsatt for a detailed scholarly account of Poe's sources (credited and otherwise). See Standage for a detailed and definitive account of how the chess-player actually worked, and the various points on which Poe's technical solution succeeded or failed.

[135] It is interesting that the image of the man/boy/dwarf hidden within the body of the Turk, reasoning through the game, pulling strings and levers in order to operate the latter's limbs, precisely reproduces the dualist image of the human being as understood by the tradition of Cartesianism. Not the ghost, as it were, but the dwarf, in the machine.

[136] Throughout the essay, Poe uses the term "pure machine" to indicate an automaton whose functions are "unconnected with human agency" (Poe "Maelzel" 321). It should be remembered that within the Cartesianism informing Poe's concept of the machine, contingency, thought, agency, will, and mind tautologically identify the human. In contrast, while the notion of the "pure machine" is almost precisely the image used by Babbage in order to distinguish his original Difference Engine from earlier, "less useful," mechanical calculators, for him the absence of human agency suggested the presence of something like a mechanical mind (Babbage *Passages* 30, passim).

[137]
> Some of these observations are intended merely to prove that the machine must be regulated by mind, and it may be thought a work of supererogation to advance farther arguments in support of what has already been fully decided. But our object is to convince, in especial, certain of our friends upon whom a train of suggestive reasoning will have more influence than the most positive *a priori* demonstration (Poe "Maelzel" 323).

[138] Interesting, as well, is the fact that his argument largely reproduces Descartes' argument concerning animals and speech. This is not insignificant or random, as it is the implicit definition of speech as a game of skill (that is, as a determined set of rules operated upon contingently by one agent in response

to another agent operating contingently under the same rules) that links chess and speech for early AI engineers like Alan Turing (Turing 433-435; Standage 225-230). See Chapter 4.

[139] This follows directly upon a litany of previously-known automata taken directly from Brewster. But, the effect achieved by Brewster, who also follows Vaucanson's duck, et al., with Babbage's machine is to draw a category distinction between the Difference Engine and previous mechanisms. The effect of Poe's reiteration of Brewster is precisely the reverse. The force of Poe's *a priori* argument instead thrusts the Difference Engine back into the list of mechanized animals and mindless bodies.

[140] These arguments should by now sound very familiar.

[141] It is significant that Poe's understanding of mathematics expressed here cannot conceive of anything like the Bernoulli numbers which describe the contingent development of determined rules, and inform Babbage's work on the Analytical Engine. We have also seen how Babbage solved the chess problem logically by dealing with each move separately, and treating the game as a unfolding series of "best" moves, rather than attempting to predict the entire game in advance.

[142] Again, note that this is an iteration of Descartes' argument regarding animals/automatons and speech. "[I]t is not conceivable that such a machine should produce different arrangements of words so as to give an appropriately meaningful answer," and it is "impossible for a machine to have enough different organs to make it act in all the contingencies of life in the way in which our reason makes us act" (Descartes *Discourse* 140).

[143] The slippage between induction and deduction in Poe's concept of "ratiocination" would be the subject of another dissertation. Indeed, numerous philosophers from Hume to Popper have challenged the possibility of achieving truth from induction, and, it is arguable that any logical operation which yields truth must be, by definition, deductive. That is, the concept of induction tends to collapse into the concept of deduction, and it can be argued that any truth is necessarily *a priori* truth, so that the boundary between the two concepts of thought is unstable. It is certainly unstable within Poe's *oeuvre*.

Poe's awareness of this limit of induction informs his admission that

> These tales of ratiocination ["Murders in the Rue Morgue" and "The Purloined Letter"] owe most of their popularity to being something in a new key. I do not mean to say they are not ingenious—but people think they are more ingenious than they are—on account of their method and air of method. . . . Where is the ingenuity of unravelling a web which you yourself (the author) have woven for the express purpose of unravelling?" (Poe letter to Philip Pendleton Cooke, August 9, 1846.)

What is consistent within the essay on the chess-player is that Poe engages the machinery of "ratiocination" in order to produce a certainty of results, and both "Maelzel's Chess-Player" and the later "ratiocination" stories are offered as demonstrations of the reliability of the method, given the competence and skill of the thinker utilizing it. (After all, Dupin is always right). That is, Poe's answer to the question "How shall we think?" is "mechanically."

[144]
> Now even granting (what should not be granted) that the movements of the Automaton Chess-Player were in themselves determinate, they would be necessarily interrupted and disarranged by the indeterminate will of his antagonist (Poe "Maelzel's" 319).

[145] As we have seen, Poe does not invent this argument.

[146] Babbage believed this as well: "if the automaton made the first move rightly, he must be able to win the game" (Babbage *Passages* 350). As Bell points out, the history of computing has demonstrated that this is not the case (Bell 5).

[147] We have already encountered this latter formulation in newspaper accounts of the Turk in the United States, and in Descartes' *Regulae*.

[148] Of course, Poe's strong advocacy for deductive, or *a priori*, reasoning is well-known, and finds its most complex expression in the tales involving Dupin and the other "mystery" stories.

[149] Poe strangely ignores the fact that he has just described this as the characteristic functioning of the pure machine--in opposition to the function of the human mind.

[150] The ironic tone of this comment is obvious.

[151] Here Poe is discussing Willis.

[152] "Maelzel's Chess-Player" is widely acknowledged by Poe scholars as a prototypical example of the "ratiocination" that will define the later mystery stories, while numerous scholars and researchers have challenged its status as an example of effective or original reasoning (Wood 79; Standage 183-184; Wimsatt 139, passim). Carroll, ignoring all literary dimensions, calls it "a remarkable piece of hack work" (Carroll 84).

[153] For which Poe feels the need to apologize.

[154] Poe believes that the "right arm of the man within is brought across his breast, and his right fingers act, without any constraint, upon the machinery in the shoulder of the figure" (Poe "Maelzel" 326). Regarding this interesting detail, as well as many others, however, Poe's solution is incorrect (Standage 184, 194-204).

[155] Again, it is not Poe's reasoning skill that is at stake, but the effect of "ratiocination" that his language produces, and the image(s) of thought that it iterates.

[156] The corollary to this is found in Poe's suggestion elsewhere that animal instinct is superior to human reason. Between the two effects, the human seems to disappear entirely (Poe "Instinct and Reason").

[157] Bierce really may have been thinking of Poe here. As we shall see, he directly satirizes Poe's worried objections to the mechanical chess-player in "Moxon's Master."

[158] The period also produced a large number of stories in which the human appears uncannily mechanistic, such as in Frederic Jesup Stimson's "Dr. Materialismus."

[159] Shelley's monster was arguably not an automaton.

[160] While I recognize that "The Bell-Tower" is an exceedingly complex story with interventions in multiple systems of politics and meaning, of necessity here, I am ignoring some of the most glaring tropes and imagery, for example those connected to American slavery, the American Civil War, and the biblical story of Babel.

[161] A further transformation of this principle, such as that effected by Alan Turing's Imitation Game, would leave us with neither human nor machine, that is, with the posthuman. (See Chapter 4.)

[162] We may read "deterministic" and "contingent" for "rigid" and "pliant."

[163] The "but" here is not insignificant, suggesting the inferiority of the human as compared to the automaton.

[164] Haman's designs for genocide against the Jews were recorded in the Book of Esther.

[165] "[H]is seclusion failed not to invest his work with more or less of that sort of mystery pertaining to the forbidden" (Melville 142). Melville repeatedly invokes the alchemist, figuring Bannadonna's science as a hidden, mysterious, and by implication, black art.

[166] Melville's use of the word domino has multiple valences, but for our purposes it is enough to note its suggestion of the automaton's mastery over his human creator.

[167] This story, originally appeared in the New York *Sun* in 1879, and features what is probably the first mention of a cyborg in the English language. The *Sun*, which had perpetrated the Great Moon Hoax of 1835, figures prominently in the related histories of science fiction and science hoaxes, and it is worth noting that this story appeared in that newspaper both unattributed and without any extra-textual signifiers of its status as fiction. While not generally considered a hoax itself, the story's content and style, as well as its publication history, would have produced a certain effect of undecidability as to the level of truth that the story was meant to express (Moskowitz ix-lxxi).

[168] The name ironically prefigures the story's ending, in which

Mr. Fisher, like the sailors on Descartes' ship, throws the mechanical brain overboard from the steamer heading back to America.

[169] Rapperschwyll is Mitchell's iteration of the stock mad scientist of early American (anti)science fiction, a Faustian character almost always identified as a materialist with some European connection and designs on ruling the world, whose own apparent lack of human sympathies makes both his materialist science and his own ontological status suspect. See Frederic Jessup Stimson's "Dr. Materialismus," a direct descendant of Mitchell's Rapperschwyll, for a particularly vivid example of this character and the anxieties that it expresses.

[170] Whether Baron Savitch actually names the cyborg or only the mechanical brain within Stépan Borovitch's skull, shifts throughout the story, and is symptomatic of the ontological uncertainty that it signifies.

[171] Even the details with which the story begins "may or may not be remembered" (Mitchell 25).

[172] Miller makes a similar observation (Miller 141).

[173] The narrator is gripped with "spiritual exaltation" upon contemplating Moxon's philosophy.

[174] This last observation effects yet another confusion of Cartesian categories, since it seems to use the image of the animal to suggest the machine's lack of reason, while at the same time playing upon an opposition between animal and machine.

[175] Elsewhere within the same volume, Bierce writes that "the most dreadful of all existences" is the "body without a soul" (Bierce "The Death of Halpin Frayser" 8).

[176] Wiener specifically attributes the ontological separation between human and automata, and between human and animal, to Descartes (Wiener *Cybernetics* 40; *God and Golem* 4).

[177] Wiener writes, "the most fruitful areas for the growth of the sciences were those which had been neglected as a no-man's land between the various established fields" (Wiener *Cybernetics* 2)--that is, in the spaces where two systems might communicate, thus drawing one another into themselves,

transforming one another toward the emergence of a new system.

[178] Wiener's research partner for this work was Julian Bigelow, and the initial work was done on the Bush differential analyzer, an analog computer (Wiener *Cybernetics* 6). It is an interesting feature of Wiener's career that he almost always worked in collaboration with other researchers.

[179] Heisenberg's principle specifically forbids the simultaneous determination of a particle in both time and space. Thus, "[i]n quantum mechanics," in contrast to Newtonian physics, "the whole past of an individual system does not determine the future of that system in any absolute way but merely the distribution of possible futures of the system" (Wiener *Cybernetics* 93). That is, prediction becomes a statistical problem.

[180] Feedback, it should be noted, understood as a function of communication, provides the basis for the central insight of cybernetics that communication is control. Thus, for instance,
> when we desire a motion to follow a given pattern the difference between the pattern and the actually performed motion is used as a new input to cause the part regulated to move [more consistently with] the pattern (Wiener *Cybernetics* 6-7).

[181] While Turing was not part of this group, Wiener credits him as "perhaps first among those who have studied the logical possibilities of the machine as an intellectual experiment" (Wiener *Cybernetics* 13, 125-126).

[182] Note especially the indeterminate referent in the following passage:
> [T]here is a human gun-pointer or gun-trainer or both coupled into the fire-control system, and acting as an essential part of it. [. . .] Moreover, their target is also humanly controlled, and it is desirable to know *its* performance characteristics (Wiener *Cybernetics* 6; emphasis mine).

[183] That is, in cybernetic terms, "interference."

[184] Simply put, this refers to the distinction between individual

learning and species learning (as, for instance, through the mechanism of DNA).

[185] A transducer is any electronic device for transforming one form of energy into another. A black box is one whose interior arrangement is not known, whereas the interior of a white box is observable.

[186] That Wiener knew of Babbage's work on computing engines is certain, as he mentions it in a discussion of Bush's differential analyzer (Wiener *Human Use* 149).

[187] Wiener is somewhat confused here about the details.

[188] When asked, in reference to his own paper chess machines, "who was learning, you or the machine?" Turing answered, "Well, both, I guess" (Turing "Can Automatic Calculating Machines Be Said to Think?" 497).

[189] "The class of problems capable of solution by the machine can be defined fairly specifically. They are those problems which can be solved by human clerical labour, *working to fixed rules*, and without understanding" (Turing "Proposed Electronic Calculator" 14; emphasis mine).

[190] Turing completed this paper while a Fellow at King's College in Cambridge, where Babbage had been Lucasian Professor of Mathematics from 1828-1839. While Turing makes no mention of this fact in "On Computable Numbers," the similarity of his ideas for the stored program, subsidiary tables (subroutines), and the universal computing machine itself, to Babbage's ideas for programmed punch cards and a "general" computing engine is remarkable.

[191] "Thus the complexity of the machine to be imitated is concentrated in the tape and does not appear in the universal machine proper in any way" (Turing "Lecture on the Automatic Computing Engine" 383).

[192] The language here is vaguely evocative of Poe's "pure machines." However, one has to wonder also whether Turing is thinking of Babbage's distinction between calculating engines "that perform the whole operation without any mental attention once the given numbers have been put into the

machine" and those "less useful" calculating machines that do "require a moderate portion of mental attention"--the former of which Babbage characterizes as "really automatic" (Babbage *Passages* 31).

[193] This, of course, is a description of binary code, and the computer using it is necessarily a digital computer (though, of course, not all digital computers use binary code).

[194] Von Neumann was also at Princeton during this time, and knew Turing well. Copeland reports that Turing's early work had a broad influence on von Neumann. Von Neumann may also have had considerable influence on Turing, particularly as game theory relates to Turing's interest in the play between (deterministic) rules and random interference (contingency) in self-organizing systems.

[195] ENIGMA was itself a portable electro-mechanical device for encoding messages that could be mechanically "programmed" or configured to produce a new code for each set-up (Copeland 220-231).

[196] In 1948, Turing wrote:
> We are then faced with the problem of finding suitable branches of thought for the machine to exercise its powers in. The following fields appear to me to have advantages:
>   (i) Various games e.g. chess, noughts and crosses, bridge, poker
>   (ii) The learning of languages
>   (iii) Translation of languages
>   (iv) Cryptology
>   (v) Mathematics ("Intelligent Machinery" 420).

[197] The original sources for material quoted in Bell are not individually cited in Bell. Frustratingly, he provides only one note at the end of the book under the heading "References" which directs the reader to the *Proceedings of the First Computer Chess Conference*--a volume for which no known copies remain extant (Hopgood). However, as Bell was among the first to develop a chess program to play at the Master level, and as a "great deal of material in this book came from personal communications" with other chess programmers and AI engineers, the volume is a valuable source despite its scholarly

flaws.

[198] Claud Elwood Shannon, like Turing, had spent a year studying mathematics and game theory under von Neumann at Princeton before World War II, and had worked on the problem of automatic code-breaking systems during the war. In 1949, Shannon presented a paper which detailed a computer chess-playing program for an electronic computer, capable of looking ahead three moves in order to evaluate, save, and implement the best course (Bell 21). A year later, he published "Programming a Computer for Playing Chess." These principles of "foresight," "memory," and evaluation, were, as we have seen, already present in Babbage's Analytical Engine, and in his less rigorous proof that "every game of skill is susceptible of being played by an automaton" (Babbage *Passages* 350).

[199] Bell quotes from Turing's notes on the game. A poor player, Turing described his paper machine as a caricature of his own weak game, making "oversights which are very similar to those which I make myself" (Turing "Chess" 574; Bell 21;). This fact not only suggests one possible reason for Turing's fascination with chess, but ironically casts some doubt on the idea of chess-playing as a straight-forward test of intelligence.

[200] Champernowne, however, gives this credit to his wife (Copeland 563).

[201] Turing had been specifically recruited to design an electronic version of a universal Turing machine by the National Physical Laboratory in 1945. The ACE (Automatic Calculating *Engine*) had been named by its director John Womersley in homage to Charles Babbage (Copeland 363). Thus, Turing would have been at least casually acquainted with the work of Charles Babbage upon joining the ACE team, but probably knew a great deal about Babbage much earlier.

[202] In an interesting iteration of Cartesian mind/body dualism, Turing writes in a memo to Womersley in 1946, referring disparagingly to von Neumann and the EDVAC project, of the "American tradition of solving one's difficulties by means of much equipment rather than thought" ("Turing's Comments on Wilkes'"). It is worth noting, however, that in this and other

passages, Turing is pointing to the relative practical differences between more or less contingent systems, not ontological or substantial differences between matter and thought. And, it is, in fact, precisely this transformation of Descartes' original problematic that connects all of Turing's work through its multiplicity of iterations.

[203] In 1946 and 1947, Turing gave a series of lectures in London on computer design. At least one of the Manchester project's chief engineers recalls attending these lectures (Copeland 373).

[204] This description of Turing's theoretical interests while on leave from ACE is quoted in a memo from the director of the NPL.

[205] The report, entitled "Intelligent Machinery" was described by Turing's superiors as "thin" and "unsuitable for publication" (Copeland 401).

[206] The "Manchester Baby," under the directorship of Newman, and based on Turing's programming models, was the first operational stored-program computer (Copeland 367).

[207] For Turing, the shift to AL was not a change of focus, but represented yet another iteration of the problematic he had been working on for years.

[208] "The Analytical Engine has no pretensions whatever to *originate* anything. It can do whatever we *know how to order it* to perform" (Lovelace 300). Turing was almost certainly thinking of this passage, as it appears as a direct quotation only a few years later in his essay in *Mind* ("Computing Machinery and Intelligence" 455). His source here may be D. R. Hartree, whom he cites in "Computing Machinery and Intelligence." However, Lovelace's comment, as Turing acknowledges in the latter essay, is in fact less pessimistic than it is insightful, as she struggles to articulate the mutually productive relationship between the deterministic programming of the Analytical Engine and its potential as "a thinking or [as] a reasoning machine" (Menabrea 373).

[209] Turing's later writing makes it clear that he understands

the human brain to be a particularly complex example of a Turing machine, as in, for example, "the whole mind is mechanical" ("Computing Machinery and Intelligence" 459), or "in so far as a man is a machine" ("Intelligent Machinery" 421). Turing also makes the point in his 1947 lecture to the London Mathematical Society:

> Some years ago I was researching [. . .] digital computing machines. I considered a type of machine which had a central mechanism, and an infinite memory which was contained on an infinite tape. [. . .] Machines such as ACE may be regarded as practical versions of this same type of machine" ("Lecture on the Automatic Computing Engine" 378).

But, the foundations of this insight are already evident when he writes in 1936 that "[w]e may compare a man in the process of computing a real number to a machine which is only capable of a finite number of conditions" ("Computable Numbers" 59). Over a decade later, he will write more definitively that a "man [. . .] is in effect a universal machine" ("Intelligent Machines" 417).

[210] Copeland reads this as equivocation, with Turing making a distinction between thinking and emotions (Copeland 566). However, this reading ignores the context of the exchange, as well as Turing's overall logic. It seems clear that rather than buying into the thought/emotion distinction suggested by the questioner, Turing is treating emotion as simply one function of thought or mind, and applying to it the same argument that he has already made regarding thought generally. That is, since we can never really know what any other agent is thinking *or* feeling, we ought to just assume that things that *appear* to be thinking or feeling *are* thinking or feeling. Thus, this passage emphasizes rather than mitigates Turing's strong belief that it is only a kind of human chauvinism that prevents us from ascribing these functions to machines.

[211] While the best-known exposition of what is now known simply as the Turing Test appeared in 1950 in "Computing Machinery and Intelligence," the idea also appeared in a variety of other forms throughout Turing's career. The original Imitation Game takes place among three people, one man (A), one woman (B), and one other person of either sex (C). The

interrogator is in a separate room, where s/he cannot see or hear the other two. Voice and other cues are eliminated by the use of a typewriter and an intermediary or a teletype machine. The interrogator may ask any question, and must try to guess which of the others is the man. A's job is to fool the interrogator into thinking he is the woman; B's job is to help the interrogator make the right choice ("Computing Machinery and Intelligence" 441).

Turing asks, "'What will happen when the machine takes the part of A in this game?' Will the interrogator decide wrongly as often [. . .] as he does when the game is played between a man and a woman?' These questions replace our original "Can machines think?'" ("Computing Machinery and Intelligence" 441).

[212] It is worth noting that Turing's treatment of language seems to anticipate many poststructuralist insights. Thus, while Descartes' human being is identified by speech-as-contingency, Turing's work stresses the contingency of written codes, and in fact, tends to erase non-functional distinctions between spoken and written language entirely (for example, the conversation at the center of the Imitation Game is written). Thus, for Turing, any kind of coding system can be more or less determined, but it is the degree of play between its ordered rules (*langue*) and its contingent unfolding (*parole*) that organizes it as a dynamic (and therefore interesting) system.

[213] Recall that Turing and others from Bletchley Park had been doing just that for years by this time.

[214] Note also that this claim is made at least three years before the historic game with Alick Glennie (Bell 17).

[215] In his essay for *Mind*, Turing playfully notes that "we wish to exclude from the machines [concerned here] men born in the usual manner" ("Computing Machinery and Intelligence" 443).

[216] Using the analogy of an atomic pile reaching critical mass to suggest the process of emergence, Turing asks, "Is there a corresponding phenomenon for minds? There does seem to be one for the human mind. [. . .] 'Can a machine be made to be super-critical?'" ("Computing Machinery and Intelligence" 459).

[217] Turing describes this as the "theological objection," where

[t]hinking is a function of man's immortal soul [and] God has given an immortal soul to every man and woman, but not to any other animal or to machines. Hence no animal or machine can think." Turing is, he says, "unable to accept any part of" this objection to the concept of a thinking-machine ("Computing Machinery and Intelligence" 449).

[218] Later, Turing does, for the purposes of the essay, define the "machine" as "an 'electronic computer' or digital computer'" (443).

[219] Shannon makes a similar observation when he suggests that "chess is generally considered to require 'thinking' for skilful play; a solution of this problem will force us either to further restrict our concept of 'thinking' or to admit the possibility of mechanised thought" (qtd. in Bell 22).

[220] This, as we have seen, is precisely the solution advanced by Babbage for the Analytical Engine, as well as his own outline for a chess program.

[221] In a 1946 newspaper interview, Turing estimates that within "about a 100 years time" the possibility of an electronic computer capable of playing an average game of chess will have been "settle[d] experimentally" (qtd. in Bell).

[222] Turing, like Wiener, would certainly have know about von Kempelen's chess-player, though I have found only one textual instance to suggest this. In a 1951 discussion of a system for machine learning which specifically references the game of chess, Turing warns against arranging the machine's experiences in order to gain a predetermined result. This, he characterizes as "a gross form of cheating, almost on a par with having a man inside the machine" (Turing "Intelligent Machinery" 473).

[223] "The class of problems capable of solution by the machine can be defined fairly specifically. They are those problems which can be solved by human clerical labour, *working to fixed rules*, and without understanding" (Turing "Proposed Electronic Calculator" 14; emphasis mine).

[224] We have already seen Bierce make this observation rather more dramatically with "Moxon's Master."

[225] Including *Mind Children*, the very book that Hayles cites.

[226] While popular and admittedly silly images of human minds being transferred into and through machines abound (*Gilligan's Island*, *Star Trek*, and *The Monkees* come first to mind), Hayles' attribution of this image to AI theorists is not entirely fair. As we have seen, Turing himself was keenly aware that the specific embodied iteration of any thinking program will affect its development (or becoming), and it is for this reason that he considered the specific shape of any particular machine intelligence to be largely unpredictable. Moreover, this principle is at the very core of learning-based emergence models of AI engineering and theory. See, for instance, Alan Turing's discussion of the problem of machine learning in the absence of human bodies or experience in "Intelligent Machinery" (420-421).

[227] "Just possibly, human personalities could participate [. . .] in the mainstream of this future activity [. . .] by being transformed into a compatible form--surely becoming very unhuman in the process" (Moravec *Robot* 12).

[228] Note that Hayles' language here is decidedly McLuhanesque.

[229] For instance, in *The Human Use of Human Beings* Wiener clarifies his concerns this way: "The thesis which I wish to maintain is neither pro- nor anti-communist but antirigidity" (84).

[230] Like Wiener, the fact that his work "naturally leads to" questions about to what degree machines might take over human roles does not escape Turing:

> those who work in connection with the ACE will be divided into its masters and servants [. . .] As time goes on the calculator will take over the functions of both masters and servants (Turing *Lecture on the Automatic Computing Machine* 392).

But this does not seem to worry him. Indeed, Turing suggests, the "real danger" is that

> the masters may be unwilling to let their jobs be stolen from them in this way [surrounding] their work with mystery and mak[ing] excuses, in well chosen gibberish whenever dangerous suggestions [a]re made (Turing

Lecture on the Automatic Computing Machine 392).
Thus, Turing like Wiener, effects a transformation of this master/slave anxiety. The highly determined system, insofar as it presents a threat to contingency and creativity, remains as a cause for anxiety; but determinacy and contingency themselves are no longer linked essentially to the machine and the human respectively.

     Moreover, because they are now free to move across the ontological or conceptual boundaries that no longer oppose the human and the machine, determinacy and contingency themselves are also no longer *essentially* opposed, but are rather relationally and mutually productive. In other words, for Turing, as for complexity theory generally, it is precisely because the function of contingency survives the transformation of the human to the posthuman that anxieties about the loss of the human as an ontological category do not.

[231] Haraway reports that she has since stopped using the term, and characterizes her recent work as an attempt to "get away from posthumanism" (Haraway "When" 140).

[232] Haraway, for instance, like Hayles, mistakenly worries that a refusal to identify categories is somehow a denial of material reality, and thus of embodied experience (Haraway "When" 151; passim).

[233] To insist that identity systems always function as closed systems is simply to recognize an *a priori* fact about the concept of identity and its genealogy. This is not to say that there cannot be interventions, disruptions, and boundary crossings, but these all have the effect of not only destabilizing specific categories and identities, but of challenging the concept of identity itself.

Bibliography

Alanen, Lilli. *Descartes's Concept of Mind*. Cambridge, MA: Harvard University Press, 2003.

Alcoff, Linda Martín. *Real Knowing: New Versions of the Coherence Theory*. Ithaca: Cornell University Press, 1996.

Allen, George. "The History of the Automaton Chess Player in America." In *The Book of the First American Chess Congress*. 420-484. New York: Rudd and Carleton, 1859.

"American Institute Fair." *Scientific American* 2 Nov. 1867: 281-282. *Making of America. 19th Century Masterfile*. Paratext. 29 Mar. 2007. <http://cdl.library.cornell.edu/cgi-bin/moa>.

"The Automaton Chess Player." *The Providence Patriot*. 11 Feb. 1826: 2. Rpt. from *New York Evening Post. Early American Newspapers, America's Historical Newspapers. NewsBank/Readex/American Antiquarian Society*. Syracuse University, Syracuse, NY. 18 Mar, 2007. < http://infoweb.newsbank.com.libezproxy2.syr.edu>.

"The Automaton Chess Player." *The Sandusky Clarion*. 27 May 1826. Rpt. from *Noah's Advocate. Newspaper Archive. Access*. 18 Mar 2007. < http://access.newspaperarchive.com.libezproxy2.syr.edu>.

"Babbage." Science Museum, London. 23 Dec. 2007. <http://www. sciencemuseum.org.uk/onlinestuff/stories/babbage.aspx>.

Babbage, Charles. *The Ninth Bridgewater Treatise*. Second ed. 1838. Dig. John van Wyhe. The Victorian Web. 18 March 2007. <http://www.victorianweb.org/science/science_texts/ bridgewater/b1.htm>.

---. *Passages from the Life of a Philosopher*. 1864. Ed. and intro. Martin Campbell-Kelly. London: Pickering and Chatto, 1994.

---. "On the Theoretical Principles of the Machinery for Calculating Tables." *Brewster's Journal of Science*. letter to David Brewster. Nov. 6, 1822. Rpt. in Morrison and

Morrison.

Baker, Gordon, and Katherine J. Morris. *Descartes' Dualism*. 1996. NY: Routledge, 2002.

Berkeley, Edmund C. *Giant Brains, or, Machines that Think*. New York: John Wiley and Sons, 1949.

Bell, Alex G. *The Machine Plays Chess?* New York: Pergamon Press, 1978.

Bierce, Ambrose. *Can Such Things Be?* (1893). Cirencester: Echo Library, 2005.

---. "The Death of Halpin Frayser." Bierce. *Can Such Things Be?*

---. "Moxon's Master." Bierce. *Can Such Things Be?*

Bowden, B. V., Ed. *Faster than Thought*. New York: Pitman Publishing, 1953.

Braidotti, Rosie. "Posthuman, All Too Human: Towards a New Process Ontology." *Theory, Culture & Society* 23.7-8 (2006): 197-2008. *SagePub*. Web. 25 August 2008 <http://tcs.sagepub.com.libezproxy2.syr.edu/content/abstract/23/7-8/197>.

---. *Transpositions: On Nomad Ethics*. Polity Press, 2006.

Brewster, David. *Letters on Natural Magic*. 1832. 7th ed. London: Wm. Tegg and Co. 1856. 8 August 2007. <http://ia331314.us.archive.org/3/items/lettersonnatmagic00brewrich/lettersonnatmagic00brewrich_djvu.txt>.

Butler, Judith. *Gender Trouble: Feminism and the Subversion of Identity*. New York: Routledge, 1990.

Buxton, L. H. D. "Charles Babbage and His Difference Engines." *Newcomen Society Transactions*. Vol. XIV. London: Courier Press, 1935.

Carol, Andy. Building a Calculating Machine Using Lego Pieces. 22 Dec. 2007 <http://acarol.woz.org/>.

Carroll, Charles Michael. *The Great Chess Automaton*. NY: Dover, 1975.

Cilliers, Paul. *Complexity and Postmodernism: Understanding Complex Systems*. London: Routledge, 1998.

Cohen, John. 1966. *Human Robots in Myth and Science.* New York: A. S. Barnes and Co., 1967.

Copeland, B. Jack, ed. *The Essential Turing: The Ideas that Gave Birth to the Computer Age.* NY: Oxford University Press, 2004.

"Curiosities of the Great Exhibition." *Scientific American.* 7 June 1862: 353-368. *Making of America.* 31 March 2007. <http://cdl.library.cornell.edu/moa>.

Deleuze, Gilles, and Félix Guattari. *What Is Philosophy?* Trans. Hugh Tomlinson and Graham Burchell. New York: Columbia University Press, 1994.

Descartes, René. *Description of the Human Body. The Philosophical Writings of Descartes, Volume I.* Trans. John Cottingham, Robert Stoothoff, Dugald Murdoch. Cambridge: Cambridge University Press, 1985.

---. *Discourse on the Method. The Philosophical Writings of Descartes, Volume I.* Trans. John Cottingham, Robert Stoothoff, Dugald Murdoch. Cambridge: Cambridge University Press, 1985.

---. *Meditations on First Philosophy and Objections and Replies. The Philosophical Writings of Descartes, Volume II.* Trans. John Cottingham, Robert Stoothoff, Dugald Murdoch. Cambridge: Cambridge University Press, 1984.

---. *The Passions of the Soul. The Philosophical Writings of Descartes, Volume I.* Trans. John Cottingham, Robert Stoothoff, Dugald Murdoch. Cambridge: Cambridge University Press, 1985.

---. *The Philosophical Writings of Descartes, Volume III: The Correspondence.* Trans. John Cottingham, Robert Stoothoff, Dugald Murdoch, and Anthony Kenny. Cambridge: Cambridge University Press, 1991.

---. *Treatise on Man. The Philosophical Writings of Descartes, Volume I.* Trans. John Cottingham, Robert Stoothoff, Dugald Murdoch. Cambridge: Cambridge University Press, 1985.

---. *The World. The Philosophical Writings of Descartes, Volume*

I. Trans. John Cottingham, Robert Stoothoff, Dugald Murdoch. Cambridge: Cambridge University Press, 1985.
"Editor's Drawer." *Harper's* April 1885: 704-714. *Making of America*. 29 March 2007. <http://cdl.library.cornell.edu/cgi-bin/moa>.
"Extracts: From Miss Sedwick's New Work." *New Hampshire Sentinal*. 28 July 1841. *Early American Newspapers, America's Historical Newspapers. NewsBank/Readex/American Antiquarian Society*. Syracuse University, Syracuse, NY. 19 Dec. 2007. <http://infoweb.newsbank.com. libezproxy2.syr.edu>.
Franklin, H. Bruce. *Future Perfect: American Science Fiction of the Nineteenth Century: An Anthology*. Revised and expanded ed. New Brunswick, NJ: Rutgers University Press, 1995.
"Freud, Sigmund. "The Uncanny." 1920, 1929. *The Norton Anthology of Theory and Criticism*. Ed. Vincent B. Leitch. Trans. Alix Strachey. W. W. Norton and Company: New York, 2001.
"From the Boston Daily Advertiser." *Providence Gazette*. 23 Oct. 1822: 1. *Early American Newspapers, America's Historical Newspapers. NewsBank/Readex/American Antiquarian Society*. Syracuse University, Syracuse, NY. 19 Dec. 2007. <http://infoweb.newsbank.com. libezproxy2.syr.edu>.
"From the London Daily Advertiser: London Application of Machinery to the Calculating and Printing of Mathematical Tables." *Providence Gazette*. 23 Oct. 1822. Rpt. from *New Monthly Magazine*. *Early American Newspapers, America's Historical Newspapers. NewsBank/Readex/American Antiquarian Society*. Syracuse University, Syracuse, NY. 19 Dec. 2007. <http://infoweb.newsbank.com.libezproxy2.syr.edu>.
Gaukroger, Stephen. *Descartes: An Intellectual Biography*. NY: Oxford University Press, 1995.
Gibson, William, and Bruce Sterling. 1991. *The Difference Engine*. New York: Bantam Books, 1992.
Goldman, William, screenplay. *The Princess Bride*. 1987.

Videodisc. MGM, 2001.

Haraway, Donna J. "Cyborg Manifesto: Science, Technology, and Socialist-Feminism in the Late Twentieth Century." *Simians, Cyborgs, and Women: The Reinvention of Nature.* Haraway. New York: Routledge, 1991: 149-181.

---. Interview by Nicholas Gane. "When We have Never Been Human, What is to Be Done?: Interview with Donna Haraway. *Theory, Culture & Society* 23.7-8 (2006): 135-158. *SagePub.* Web. 25 August 2008 <http://tcs.sagepub.com.libezproxy2.syr.edu/content/abstract/23/7-8/135>.

Hayles, N. Katherine. *Chaos and Order: Complex Dynamics in Literature and Science.* Chicago: University of Chicago Press, 1991.

---. *How we Became Posthuman: Virtual bodies in Cybernetics, Literature, and Informatics.* Chicago: University of Chicago Press, 1999.

---. "Unfinished Work: From Cyborg to Cognisphere." *Theory, Culture & Society* 23.7-8 (2006): 159-166. *SagePub.* Web. 25 August 2008 <http://tcs.sagepub.com.libezproxy2.syr.edu/content/abstract/23/7-8/159>.

Haugeland, John. *Artificial Intelligence: The Very Idea.* Cambridge, Mass: MIT Press, 1985.

Heisenberg, Werner. Center for History of Physics. 14 January 2008. <http://www.aip.org/history/heisenberg/>.

Hobbes, Thomas. *Elements of Philosophy.* 1655. *English Works of Thomas Hobbes of Malmsebury.* Vol. I. Trans. and ed. Sir William Molesworth. London: John Bohn, 1839.

Hopgood, Bob. Re: FW: Proceedings of the First Computer Chess Conference. E-mail to the author. 21 March 2008.

Hyman, Anthony. *Charles Babbage: Pioneer of the Computer.* 1982. Princeton, NJ: Princeton University Press, 1983.

---, ed. *Science and Reform: Selected Works of Charles Babbage.* New York: Cambridge University Press, 1989.

Jentsch, Ernst. "On the Psychology of the Uncanny." 1906.

Trans. Roy Sellers. *Angelaki* 2.1 (1996): 7-16. *The Unheimlich.* By Matt Roberts. Web. 9 Sept. 2008. <http://www.theunheimlich.blogspot.com>.

Jesseph, Douglas. "Mechanism, Skepticism, and Witchcraft." Smaltz. NY: Routledge, 2005.

Kauffman, Stuart A. *The Origins of Order: Self-Organization and Selection in Evolution.* New York: Oxford University Press, 1991.

La Mettrie, Julien Offray de. *Machine Man and Other Writings.* Ed. and trans. Ann Thompson. Cambridge: Cambridge University Press, 1996.

Lardner, Dionysius. "Babbage's Calculating Engine." *Edinburgh Review.* July 1834: CXX. Rpt. in Morrison and Morrison.

Laszlo, Ervin. *The Systems View of the World: A Holistic Vision for Our Time.* Cresskill, NJ: Hampton Press, 1996.

Lovelace, Ada. Notes. *Sketch of the Analytical Engine Invented by Charles Babbage, Esq.* By L. F. Menabrea. Trans. Ada Lovelace. *Scientific Memoirs,* iii, 666-731. 1843. Rpt. in Hyman, *Science and Reform.* 242-312.

McCorduck, Pamela. *Machines Who Think.* 25th Anniversary Update. Natick, Massachusetts: A. K. Peters, Ltd., 2004.

McLuhan, Marshall. *Understanding Media: The Extensions of Man.* 1964. Cambridge, MA: MIT Press, 2002.

Masani, Pesi Rustom. *Norbert Wiener. Vita Mathematica, vol. 5.* Birkhäuser, 1990.

"Mechanic's Department." *New York Weekly Messenger.* 13 Feb. 1833. *Vintage Computer.* 21 Jan. 2008. <http://www.vintagecomputer.net/>.

Melville, Herman. "The Bell-Tower." *Putnam's.* August 1855: 123-130. Rpt. in Franklin.

Menabrea, L. F. *Sketch of the Analytical Engine Invented by Charles Babbage, Esq.* Trans. Ada Lovelace. *Scientific Memoirs,* iii, 666-731. 1842; 1849. Rpt. in Hyman, *Science and Reform.* 242-312.

Miller, Arthur M. "The Influence of Edgar Allan Poe on Ambrose Bierce." *American Literature* 4.2 (1932): 130-150. JSTOR

Archive. Web.

Minsky, Marvin. *The Society of Mind*. New York: Simon and Schuster, 1986.

"Miracles." *The Syracuse Herald*, Magazine Section 25 June 1922: 1-2. *Making of America*. 29 March 2007. < http://access.newspaperarchive.com.libezproxy2.syr.edu>.

"Miscellany: The Calculating-Machine." *Salem Gazette*. 16 Nov. 1832. *Early American Newspapers, America's Historical Newspapers. NewsBank/Readex/American Antiquarian Society*. Syracuse University, Syracuse, NY. 19 Dec. 2007. <http://infoweb.newsbank.com.libezproxy2.syr.edu>.

Mitchell, Edward Page. "The Ablest Man in the World." *The Crystal Man: Landmark Science Fiction*. Comp. and Intro. Sam Moskowitz. Garden City, New York: Doubleday & Company, Inc. 1973.

"More American Ingenuity." *New-Hampshire Patriot*. 5 April 1835: 1. *Early American Newspapers, America's Historical Newspapers. NewsBank/Readex/American Antiquarian Society*. Syracuse University, Syracuse, NY. 22 March 2008. <http://infoweb.newsbank.com.libezproxy2.syr.edu>.

Morrison, Philip, and Emily Morrison. Eds. Introduction. *Charles Babbage and His Calculating Engines: Selected Writings by Charles Babbage and Others*. New York: Dover, 1961.

Mitchell, Edward Page. *The Crystal Man: Landmark Science Fiction*. Comp. and Intro. Sam Moskowitz. Garden City, New York: Doubleday & Company, Inc. 1973.

Moravec, Hans P. *Mind Children: The Future of Robot and Human Intelligence*. Cambridge: Harvard University Press, 1988.

---. *Robot: Mere Machine to Transcendent Mind*. Oxford: Oxford University Press, 1999. *Google Book Search*. Web. 31 August 2008.

Moskowitz, Sam. Introduction: Lost Giant of American Science Fiction: A Biographical Perspective. Mitchell ix-lxxii.

"M. Maelzel's Automata." *Rhode Island Republican*. 26 July 1827: 4. *Early American Newspapers, America's Historical Newspapers. NewsBank/Readex/American Antiquarian*

Society. Syracuse University, Syracuse, NY. 22 March 2008. <http://infoweb.newsbank.com.libezproxy2.syr.edu>.

"Mr. Maelzel." *Baltimore Patriot*. 5 August 1829: 2. *Early American Newspapers, America's Historical Newspapers. NewsBank/Readex/American Antiquarian Society*. Syracuse University, Syracuse, NY. 22 March 2008. <http://infoweb.newsbank.com.libezproxy2.syr.edu>.

"Mr. Webster's Report." *Baltimore Patriot*. 13 Feb. 1834: 2. *Early American Newspapers, America's Historical Newspapers. NewsBank/Readex/American Antiquarian Society*. Syracuse University, Syracuse, NY. 19 Dec. 2007. <http://infoweb.newsbank.com.libezproxy2.syr.edu>.

"New London Monthly Magazine." 29 Nov. 1822: 3. *Early American Newspapers, America's Historical Newspapers. NewsBank/Readex/American Antiquarian Society*. Syracuse University, Syracuse, NY. 19 Dec. 2007. <http://infoweb.newsbank.com.libezproxy2.syr.edu>.

Parfit, Derek. *Reasons and Persons*. Oxford: Clarendon Press, 1984.

Pepperell, Robert. *The Posthuman Condition: Consciousness Beyond the Brain*. 1995. Portland, OR: Intellect, 2003.

"Perpetual Motion." *New-Bedford Mercury*. 14 Nov. 1834: 1. *Early American Newspapers, America's Historical Newspapers. NewsBank/Readex/American Antiquarian Society*. Syracuse University, Syracuse, NY. 19 Dec. 2007. <http://infoweb.newsbank.com.libezproxy2.syr.edu>.

Poe, Edgar Allan. "Instinct Vs Reason--A Black Cat." Alexander's Weekly Messenger. 29 Jan. 1840: 2, cols. 6-7. 11 March 2008. <http://www.eapoe.org/works/ESSAYS/IVRBCATA.HTM>.

---. Letter to Philip P. Cooke. 9 August 9 1846. E. A. Poe Society. 15 March 2008. <http://www.eapoe.org/ works/LETTERS/P4608090.HTM>.

---. "Maelzel's Chess-Player." *Southern Literary Messenger*. April 1836: 318-326. E. A. Poe Society. 1 April 2007. <http://www.eapoe.org/works/essays/ maelzel.htm>.

"Recreations of a Philosopher." *Harper's New Monthly Magazine.* Dec. 1864: 34-39. *Making of America.* 31 March 2007. <http://cdl.library.cornell.edu/moa>.

Robinson, Tim. *Meccano Computing Machinery Web Site: Analytical Engine.* 22 Dec. 2007. <http://www.meccano.us/analytical_engine/index.html>.

Rosenfield, Leonora Cohen. *From Beast-Machine to Man-Machine.* New York: Oxford University Press, 1940.

Ryle, Gilbert. *The Concept of Mind.* 1949. Chicago: University of Chicago Press, 2000.

Schmaltz, Tad M. *Receptions of Descartes: Cartesianisms and Anti-Cartesianisms in Early Modern Europe.* NY: Routledge, 2005.

Shannon. E. C. "Programming a Computer for Playing Chess." *Philosophical Magazine.* 41 (1950): 256-275.

Smith, Nathan D., and Jason P. Taylor. *Descartes and Cartesianism.* Newcastle: Cambridge Scholars Press, 2005.

Standage, Tom. *The Turk: The Life and Times of the Famous Eighteenth-Century Chess-Playing Machine.* New York: Berkley Books, 2002.

Stimson, Frederic Jessup. "Dr. Materialismus." *Scribner's.* Nov. 1890. Rptd. in Franklin.

Swade, Doron. *The Difference Engine: Charles Babbage and the Quest to Build the First Computer.* Rpt. *The Cogwheel Brain.* 2000. New York: Penguin-Putnam, 2002.

---. "The Shocking Truth about Babbage and His Calculating Engines." *Resurrection: The Bulletin of the Computer Conservation Society.* 32 (2004). 23 Dec. 2007. <http://www.cs.man.ac.uk/CCS/res/res32.htm>.

"To the Lumbermen of New Hampshire." *New Hampshire Sentinal.* 6 Feb. 1836: 4. *Early American Newspapers, America's Historical Newspapers. NewsBank/Readex/American Antiquarian Society.* Syracuse University, Syracuse, NY. 22 March 2008. <http://infoweb.newsbank.com.libezproxy2.syr.edu>.

Thompson, Ann. Introduction. La Mettrie ix-xxviii.

Turing, A. M. (Alan). "Chess." *Faster than Thought*. 1953. Copeland 569-575.
---. "Computing Machinery and Intelligence." *Mind* 49:236 (1950): 433-460. Copeland 441-464.
---. "Intelligent Machinery." Copeland 410-432.
---. "Intelligent Machinery, A Heretical Theory." Copeland 472-475.
---. "Lecture on the Automatic Computing Engine." 1947. Copeland 378-394.
---. "Letter from Turing to W. Ross Ashby." c. 1946. 27 April 2008. <http://www.AlanTuring.net/turing_ashby>.
---. "Proposed Electronic Calculator." Undated document 1945. The Turing Archive for the History of Computing. 27 April 2008. <http://www.AlanTuring.net/proposed_electronic_ calculator>.
---. "Turing's Comments on Wilkes' Proposal for a Pilot Machine." Undated memo 1946. The Turing Archive for the History of Computing. 27 April 2008. <http:// www.cs.usfca.edu/www.AlanTuring.net/turing_archive/archive/l/l42/L42-010.html>.
Turing, A. M., Richard Braithwaite, Geoffrey Jefferson, Max Newman. Transcript, BBC broadcast. 10 Jan. 1952. "Can Automatic Calculating Machines Be Said to Think?" Copeland 494-506.
Turing, Sara. *Alan M. Turing*. Cambridge: W. Heffer, 1959.
Walker, George. "The Chess Automaton." 1839. Rpt. in *Chess and Chess Players: Consisting of Original Stories and Sketches*. London: Charles J. Skeet, 1850: 1-38. 27 Mar. 2008. <http://books.google.com/ books?id=MfUIAAAAQAAJ&pg=PA1&lpg=PA1&dq=george+walker+%22the+chess+automaton%22&source=web&ots=nWeVNu0t3h&sig=mkRMKF57YqvRBfhsuhy2D8LAHgQ&hl=en#PPP1,M1>.
"What Are We Going to Make?" *Atlantic Monthly*. June 1858: 90-101. *Making of America*. 29 March 2007. <http://cdl.library.cornell.edu/cgi-bin/moa>.

Wiener, Norbert. *Cybernetics: or, Control and Communication in the Animal and the Machine.* 1948. 2nd. ed. 1961. Cambridge, MA: MIT, 1991.

---. *Ex-Prodigy.*

---. *I Am a Mathematician.*

---. *God and Golem, Inc.: A Comment on Certain Points Where Cybernetics Impinges on Religion.* 1964. Cambridge, Mass.: MIT Press, 1966.

---. *The Human Use of Human Beings: Cybernetics and Society.* 1950. Boston: Houghton Mifflin, 1954. Da Capo Press.

Wimsatt, W. K., Jr. "Poe and the Chess Automaton." American Literature, May, 1939:138-151. JSTOR. Syracuse University, Syracuse, NY. 13 Mar. 2007. <http://www.jstor.org.libezproxy2.syr.edu>.

Windisch, Carl Gottlieb von. *Inanimate Reason; or a Circumstantial Account of That Astonishing Piece of Mechanism, M de Kempelen's Chess-Player.* London: 1784.

Wood, Gabby. *Edison's Eve: A Magical History of the Quest for Mechanical Life.* New York: Anchor, 2002.

"The World's Progress: From a Lecture, by Professor Draper of Hampden, Sydney College." *Waldo Patriot.* 2 Nov. 1838: 1. *Early American Newspapers, America's Historical Newspapers. NewsBank/Readex/American Antiquarian Society.* Syracuse University, Syracuse, NY. 19 Dec. 2007. <http://infoweb.newsbank.com. libezproxy2.syr.edu>.

www.ingramcontent.com/pod-product-compliance
Lightning Source LLC
Chambersburg PA
CBHW052350220526
45465CB00003BA/1038